Succeed

Eureka Math®

Grade 5
Modules 5 & 6

Published by Great Minds®.

Printed in the U.S.A.
This book may be purchased from the publisher at eureka-math.org.
10 9 8 7 6 5 4 3

ISBN 978-1-64054-043-9

G5-M5-M6-S-06.2018

Learn ♦ Practice ♦ Succeed

Eureka Math® student materials for *A Story of Units®* (K–5) are available in the *Learn, Practice, Succeed* trio. This series supports differentiation and remediation while keeping student materials organized and accessible. Educators will find that the *Learn, Practice,* and *Succeed* series also offers coherent—and therefore, more effective—resources for Response to Intervention (RTI), extra practice, and summer learning.

Learn

Eureka Math Learn serves as a student's in-class companion where they show their thinking, share what they know, and watch their knowledge build every day. *Learn* assembles the daily classwork—Application Problems, Exit Tickets, Problem Sets, templates—in an easily stored and navigated volume.

Practice

Each *Eureka Math* lesson begins with a series of energetic, joyous fluency activities, including those found in *Eureka Math Practice.* Students who are fluent in their math facts can master more material more deeply. With *Practice,* students build competence in newly acquired skills and reinforce previous learning in preparation for the next lesson.

Together, *Learn* and *Practice* provide all the print materials students will use for their core math instruction.

Succeed

Eureka Math Succeed enables students to work individually toward mastery. These additional problem sets align lesson by lesson with classroom instruction, making them ideal for use as homework or extra practice. Each problem set is accompanied by a Homework Helper, a set of worked examples that illustrate how to solve similar problems.

Teachers and tutors can use *Succeed* books from prior grade levels as curriculum-consistent tools for filling gaps in foundational knowledge. Students will thrive and progress more quickly as familiar models facilitate connections to their current grade-level content.

Students, families, and educators:

Thank you for being part of the *Eureka Math®* community, where we celebrate the joy, wonder, and thrill of mathematics.

Nothing beats the satisfaction of success—the more competent students become, the greater their motivation and engagement. The *Eureka Math Succeed* book provides the guidance and extra practice students need to shore up foundational knowledge and build mastery with new material.

What is in the Succeed *book?*

Eureka Math Succeed books deliver supported practice sets that parallel the lessons of *A Story of Units®*. Each *Succeed* lesson begins with a set of worked examples, called *Homework Helpers*, that illustrate the modeling and reasoning the curriculum uses to build understanding. Next, students receive scaffolded practice through a series of problems carefully sequenced to begin from a place of confidence and add incremental complexity.

How should Succeed *be used?*

The collection of *Succeed* books can be used as differentiated instruction, practice, homework, or intervention. When coupled with *Affirm®*, *Eureka Math*'s digital assessment system, *Succeed* lessons enable educators to give targeted practice and to assess student progress. *Succeed*'s perfect alignment with the mathematical models and language used across *A Story of Units* ensures that students feel the connections and relevance to their daily instruction, whether they are working on foundational skills or getting extra practice on the current topic.

Where can I learn more about Eureka Math *resources?*

The Great Minds® team is committed to supporting students, families, and educators with an ever-growing library of resources, available at eureka-math.org. The website also offers inspiring stories of success in the *Eureka Math* community. Share your insights and accomplishments with fellow users by becoming a *Eureka Math* Champion.

Best wishes for a year filled with Eureka moments!

Jill Diniz
Director of Mathematics
Great Minds

Contents

Module 5: Addition and Multiplication with Volume and Area

Module 6: Problem Solving with the Coordinate Plane

Topic F: The Years in Review: A Reflection on A Story of Units

Grade 5
Module 5

1. The following solids are made up of 1 cm cubes. Find the total volume of each figure, and write it in the chart below.

a. 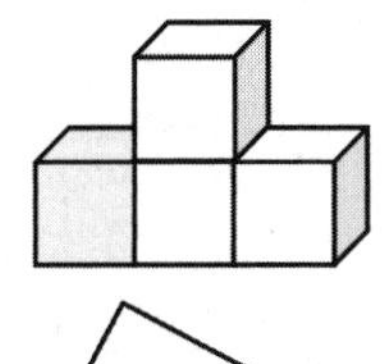

b. 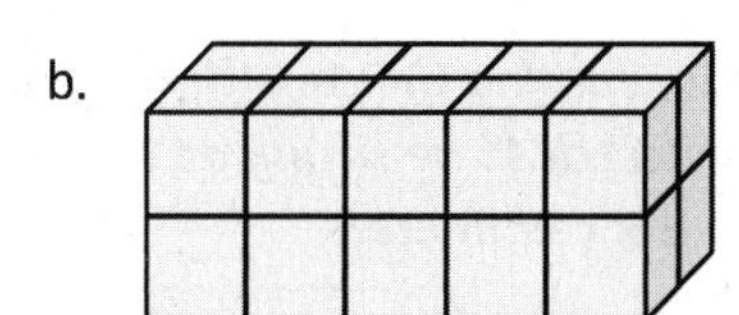

I see there are 3 cubes on the bottom and 1 cube on top. Therefore, this solid has a total of 4 cubes.

I see there are 2 layers of cubes like layers of a cake (top and bottom). There are 10 cubes on the top, and there must be 10 cubes on the bottom. Therefore, this solid has a total of 20 cubes.

Since Figure (a) is made of a total of 4 cubes, I can say that it has a volume of 4 cubic centimeters.

Figure	Volume	Explanation
a	4 cm^3	***I added 3 cubes and 1 cube.*** $3 + 1 = 4$
b	20 cm^3	***I counted the top layer and then multiplied by 2.***

2. Draw a figure with the given volume on the dot paper.

a. 2 cubic units

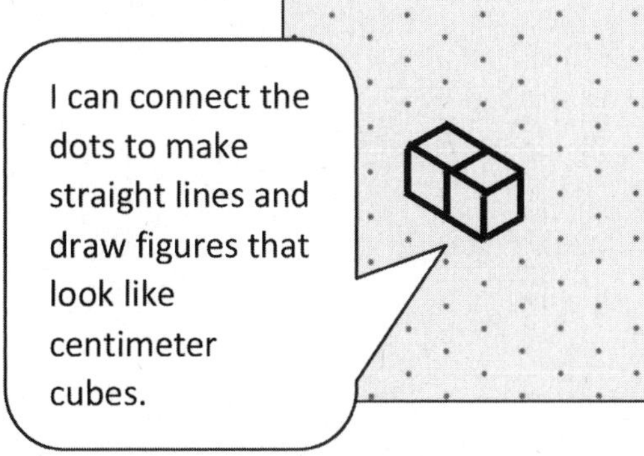

b. 4 cubic units

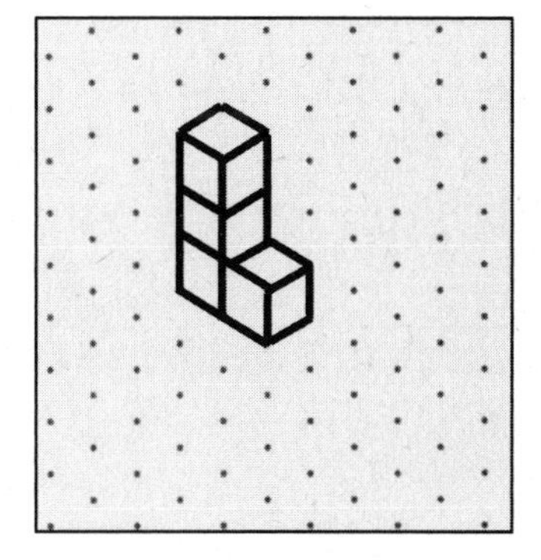

3. Allison says that the figure below, made of 1 cm cubes, has a volume of 4 cubic centimeters.

   a. Explain her mistake.

   Allison is not counting the cube that is hidden. The cube that is on the second layer needs to be sitting on a hidden cube. The volume of this figure is 5 cubic centimeters.

   I see there are 4 cubes showing, but there is one hidden under the 1 cube on top.

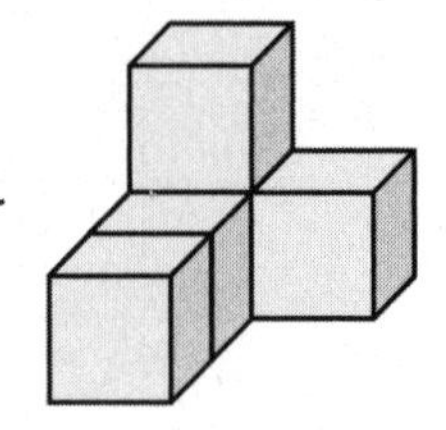

   b. Imagine if Allison adds to the second layer so the cubes completely cover the first layer in the figure above. What would be the volume of the new structure? Explain how you know.

   The volume would be $8\ \text{cm}^3$. I counted the first layer, and then multiplied by 2.

   $\mathbf{4\ cm^3 \times 2 = 8\ cm^3}$

   Since Allison wants to build a second layer that is the same as the first layer, I can just multiply 4 cubes times 2.

Name ______________________________ Date ______________

1. The following solids are made up of 1 cm cubes. Find the total volume of each figure, and write it in the chart below.

A. 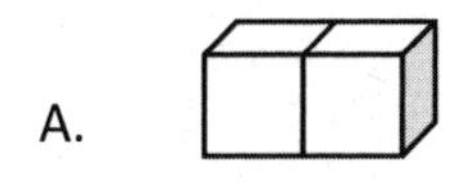

D. 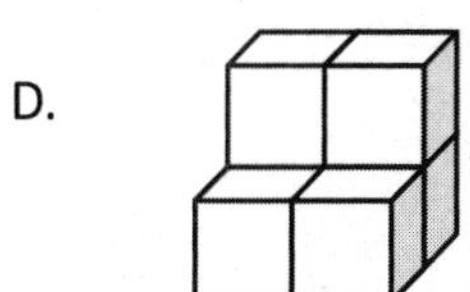

B. 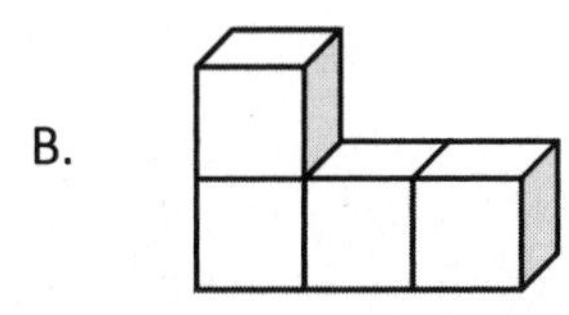

E. 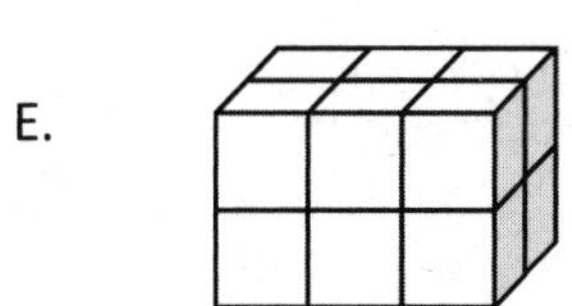

C. 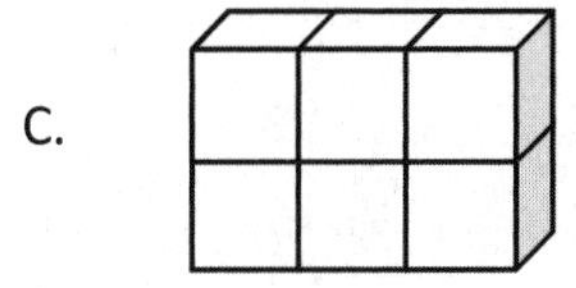

F. 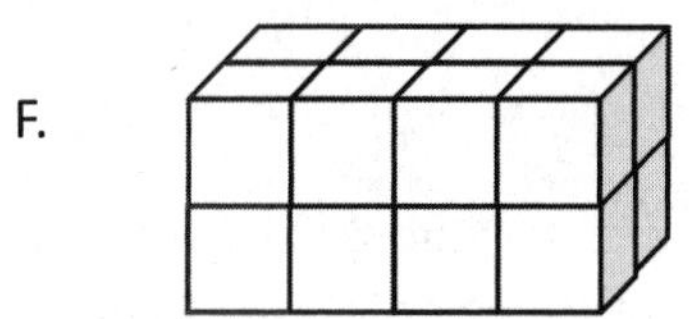

Figure	Volume	Explanation
A		
B		
C		
D		
E		
F		

2. Draw a figure with the given volume on the dot paper.

a. 3 cubic units

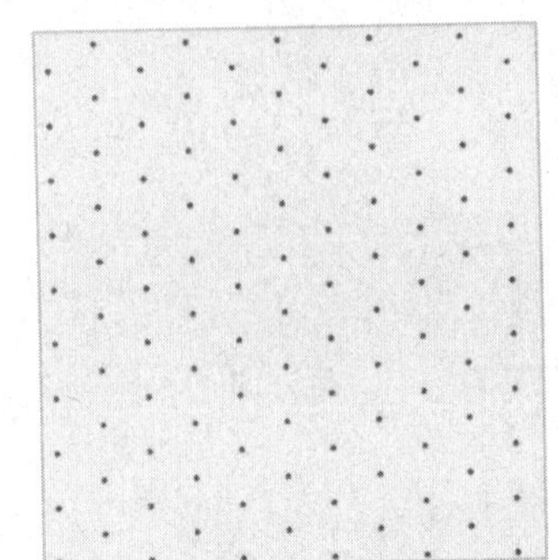

b. 6 cubic units

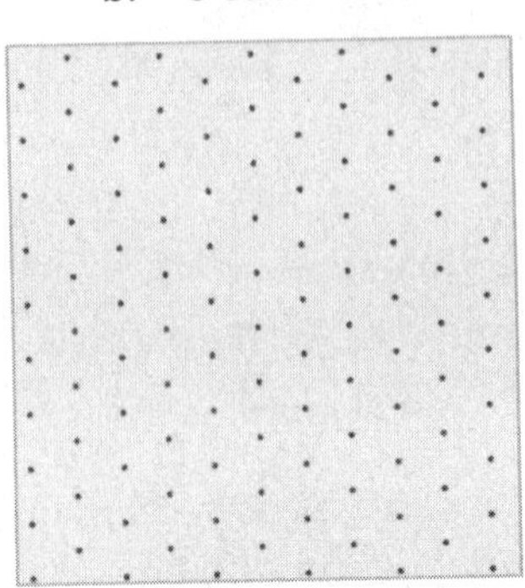

c. 12 cubic units

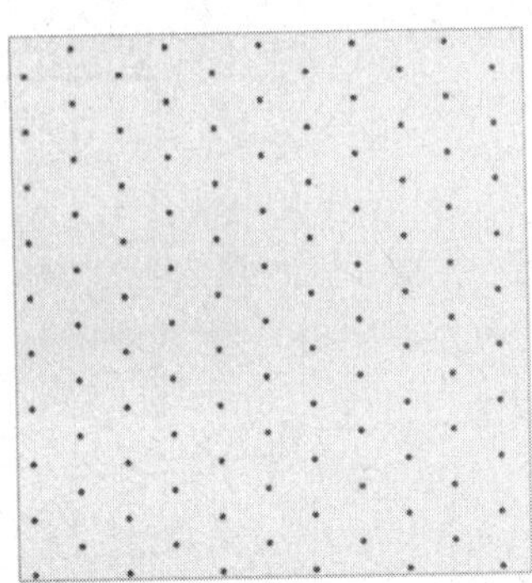

3. John built and drew a structure that has a volume of 5 cubic centimeters. His little brother tells him he made a mistake because he only drew 4 cubes. Help John explain to his brother why his drawing is accurate.

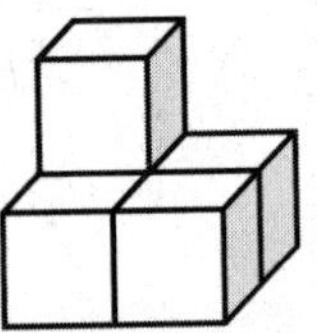

4. Draw another figure below that represents a structure with a volume of 5 cubic centimeters.

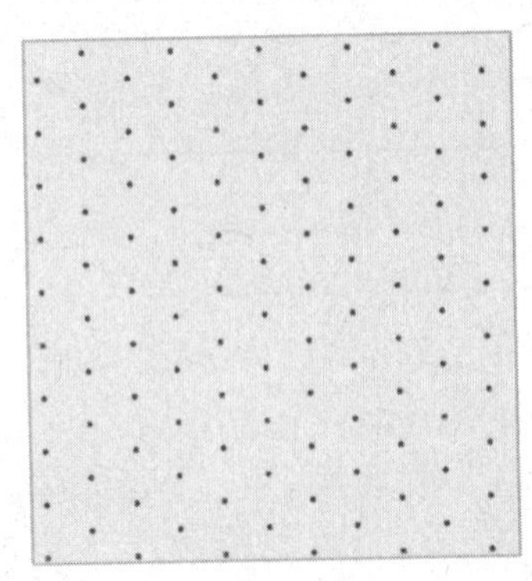

1. Shade the following figures on centimeter grid paper. Cut and fold each to make 3 open boxes, taping them so they hold their shapes. Pack each box with cubes. Write how many cubes fill the box.

a.

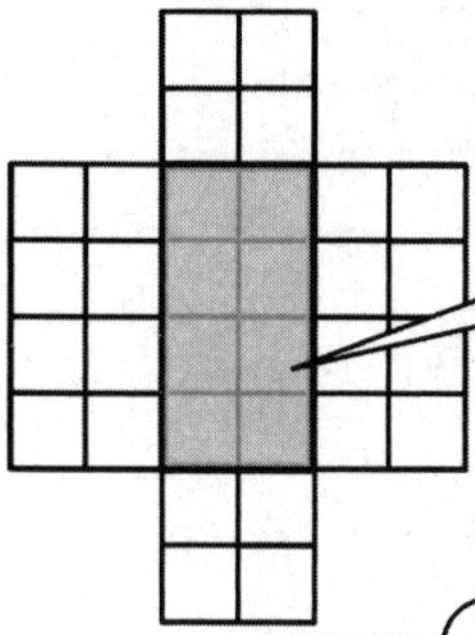

I can count the shaded area or the base. It would take 8 cubes to cover the base.

Number of cubes: **16**

I can imagine folding all of the flaps up to form an open rectangular prism. There are 2 layers (top and bottom), so I can multiply $8 \times 2 = 16$.

b.

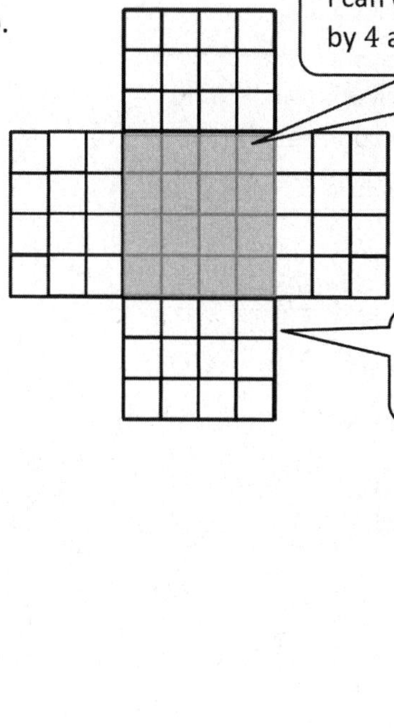

I can count the shaded area or the base. It is a 4 by 4 array, and $4 \times 4 = 16$.

Number of cubes: **48**

I can imagine folding all of the flaps up to form an open rectangular prism. There are 3 layers, so I multiply $16 \times 3 = 48$.

2. How many centimeter cubes would fit in each box? Explain your answer using words and diagrams on the box. (The figures are not drawn to scale.)

   a.

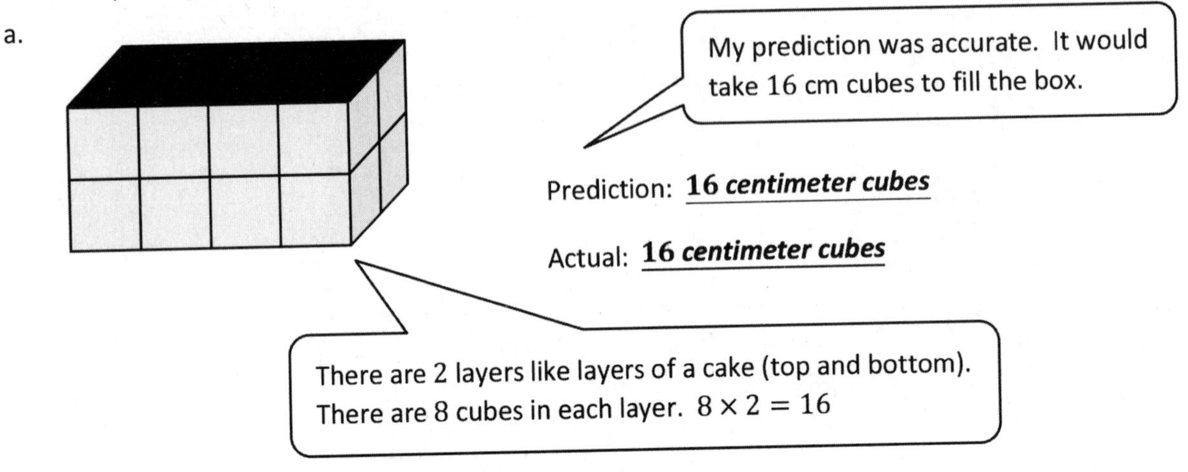

There are 2 *layers: top and bottom. Each layer has* 8 *cubes, and* 8 *cubes* $\times 2 = 16$ *cubes.*

   b.

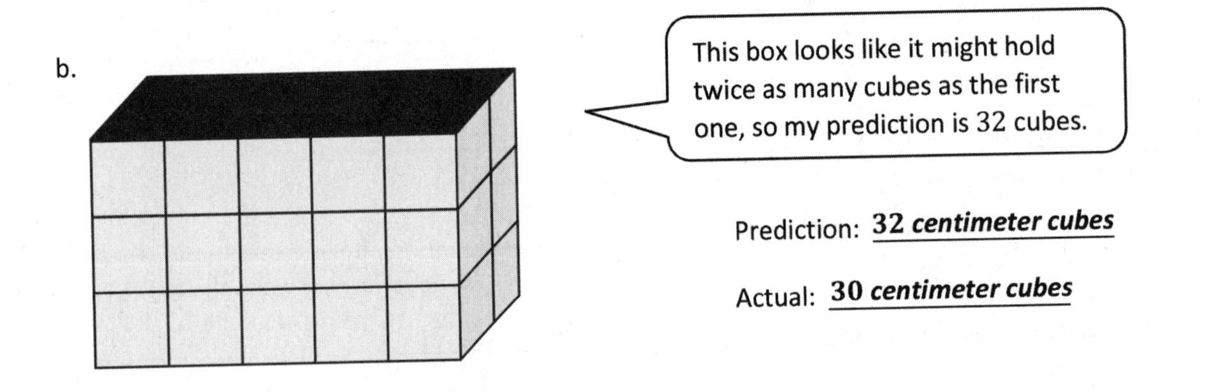

There are 3 *layers: top, middle, and bottom.*

Each layer has 10 *cubes, and* 10 *cubes* $\times 3 = 30$ *cubes.*

Name ______________________________ Date ______________

1. Make the following boxes on centimeter grid paper. Cut and fold each to make 3 open boxes, taping them so they hold their shapes. How many cubes would fill each box? Explain how you found the number.

   a. Number of cubes: ______________

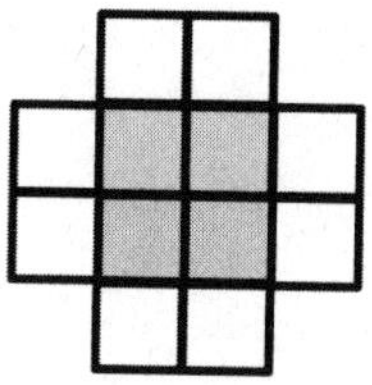

   b. Number of cubes: ______________

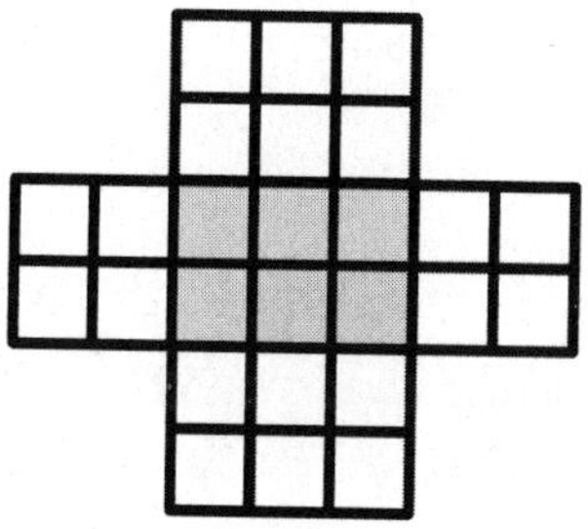

   c. Number of cubes: ______________

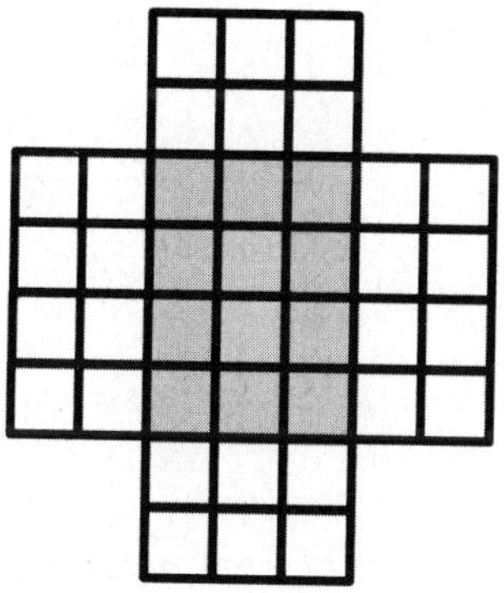

2. How many centimeter cubes would fit inside each box? Explain your answer using words and diagrams on each box. (The figures are not drawn to scale.)

a.

Number of cubes: ____________

Explanation:

b.
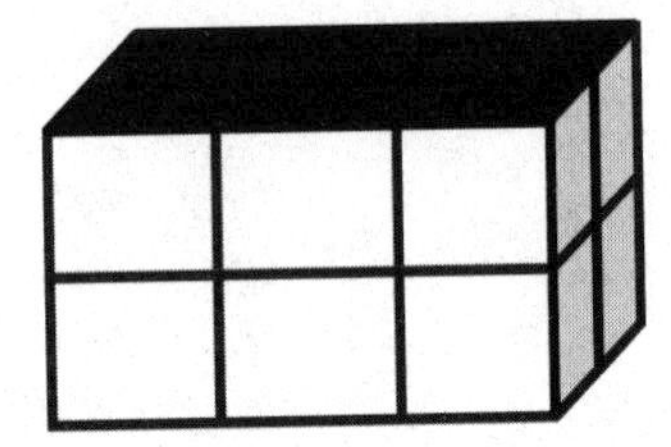

Number of cubes: ____________

Explanation:

c.
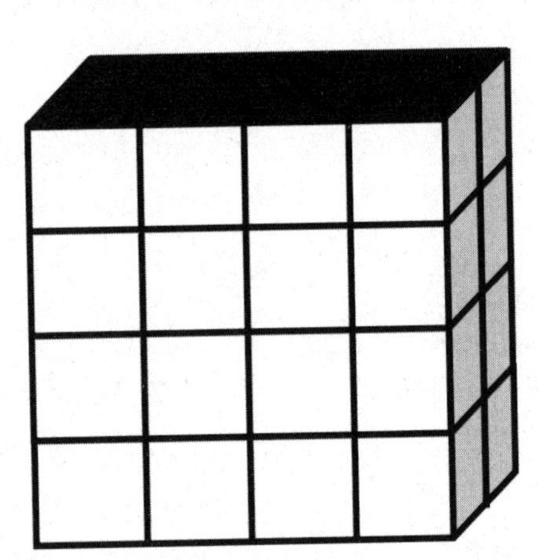

Number of cubes: ____________

Explanation:

3. The box pattern below holds 24 1-centimeter cubes. Draw two different box patterns that would hold the same number of cubes.

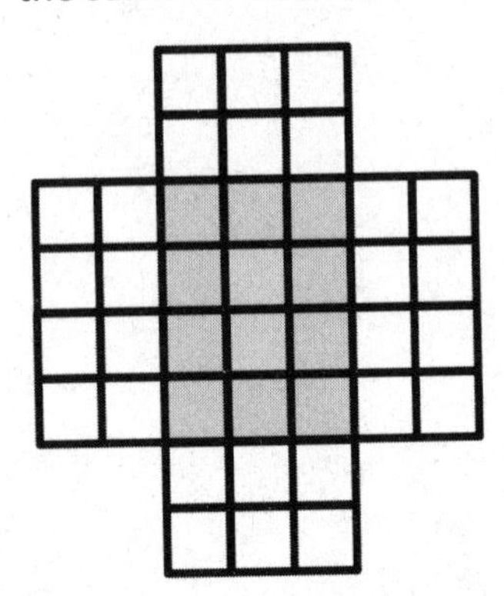

1. Use the prisms to find the volume.
   - Build the rectangular prism pictured below to the left with your cubes, if necessary.
   - Decompose it into layers in three different ways, and show your thinking on the blank prisms.
   - Complete the missing information in the table.

Number of Layers	Number of Cubes in Each Layer	Volume of the Prism
2	10	20 cubic cm
5	4	20 cubic cm
2	10	20 cubic cm

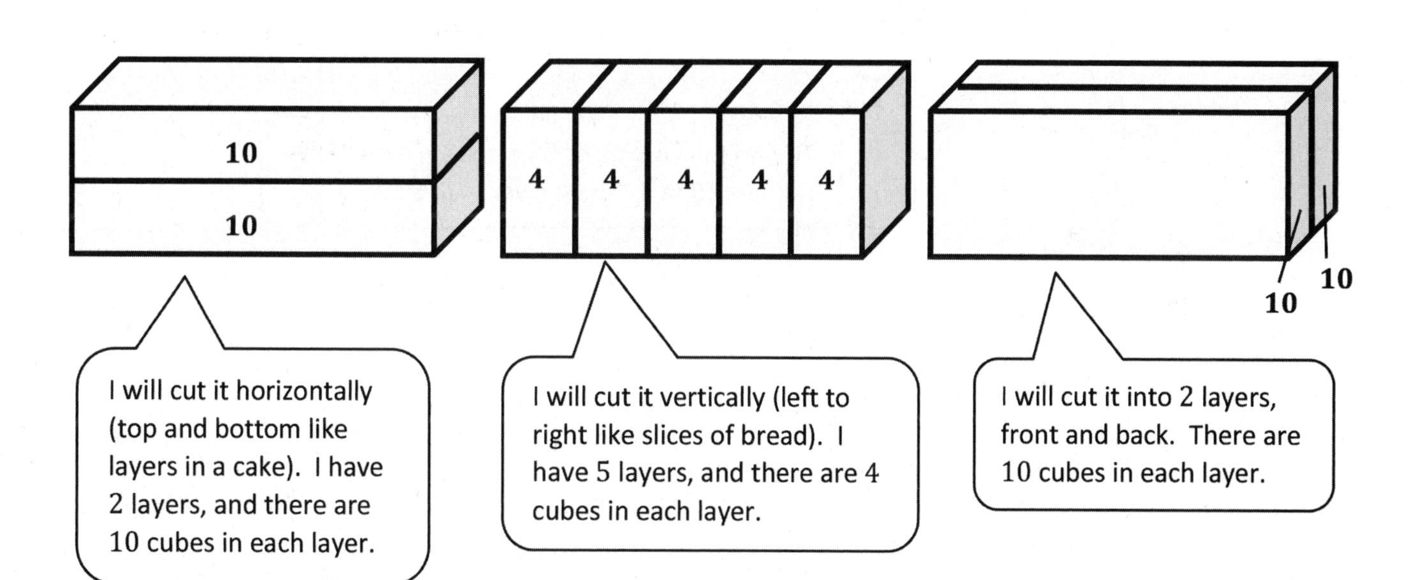

I can visualize a prism that is $5 \text{ in} \times 5 \text{ in} \times 1 \text{ in}$. When looking at the prism from the top, it would look like a square since the length and the width are equal. The prism is also just one inch tall, so it looks like the bottom layer of a cake.

2. Joseph makes a rectangular prism 5 inches by 5 inches by 1 inch. He then decides to create layers equal to his first one. Fill in the chart below, and explain how you know the volume of each new prism.

To find the volume in 3 layers, I will multiply 3 times 25 in^3. The answer is 75 in^3.

Number of Layers	Volume	Explanation
3	$\mathbf{75 \text{ in}^3}$	**1 *layer:*** $\mathbf{25 \text{ in}^3}$ **3 *layers:*** $\mathbf{3 \times 25 \text{ in}^3 = 75 \text{ in}^3}$
5	$\mathbf{125 \text{ in}^3}$	**1 *layer:*** $\mathbf{25 \text{ in}^3}$ **5 *layers:*** $\mathbf{5 \times 25 \text{ in}^3 = 125 \text{ in}^3}$

To find the volume of 5 layers, I will multiply 5 times 25 in^3. The answer is 125 in^3.

Name ______________________________ Date ______________

1. Use the prisms to find the volume.

   - The rectangular prisms pictured below were constructed with 1 cm cubes.
   - Decompose each prism into layers in three different ways, and show your thinking on the blank prisms.
   - Complete each table.

a.

Number of Layers	Number of Cubes in Each Layer	Volume of the Prism
		cubic cm
		cubic cm
		cubic cm

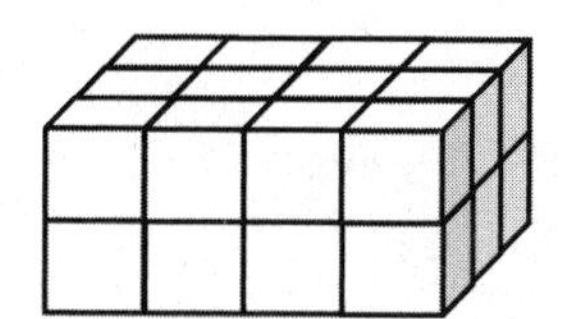

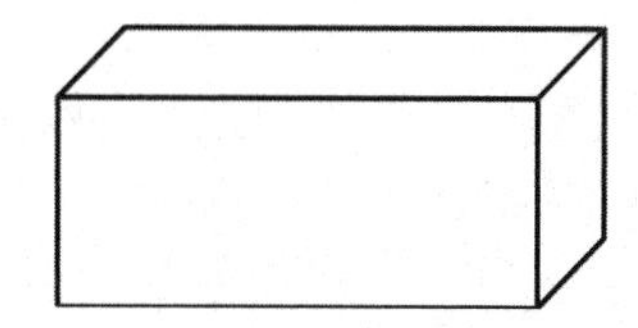
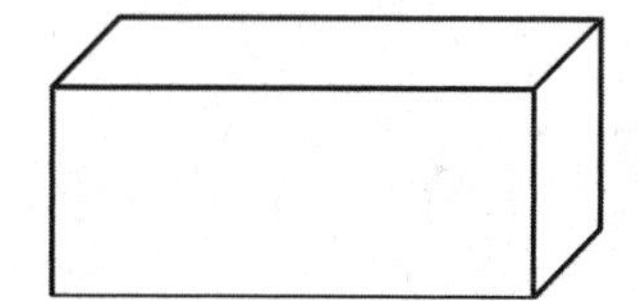
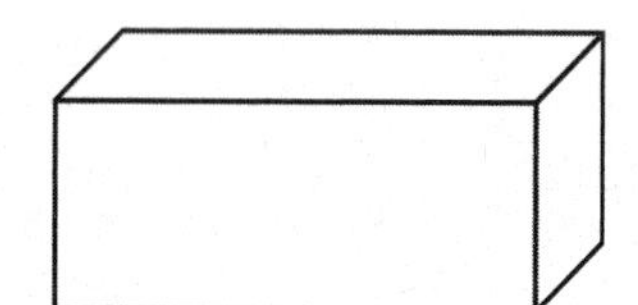

b.

Number of Layers	Number of Cubes in Each Layer	Volume of the Prism
		cubic cm
		cubic cm
		cubic cm

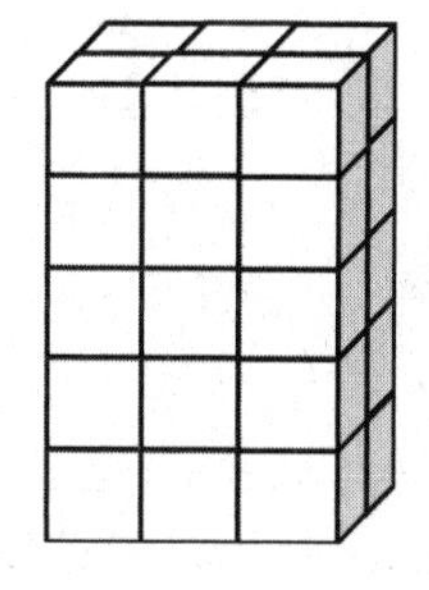

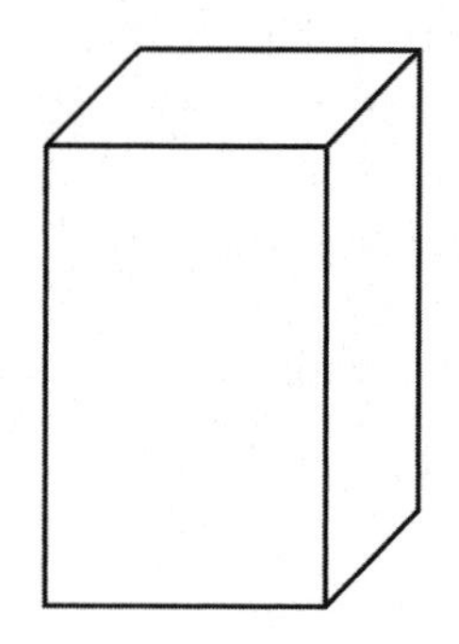
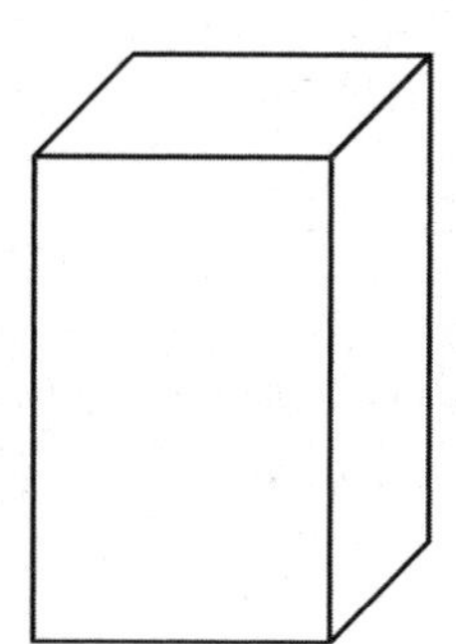
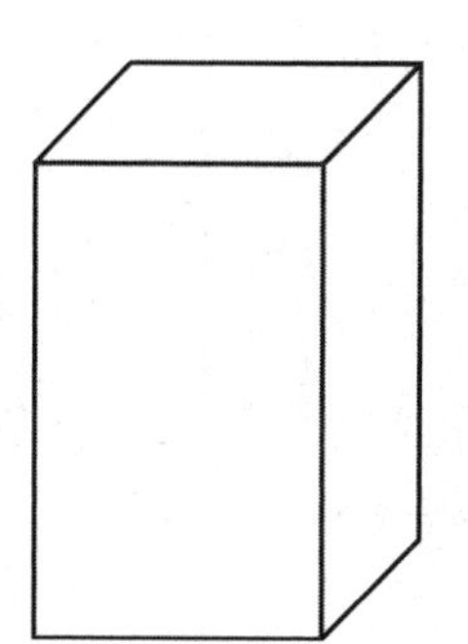

2. Stephen and Chelsea want to increase the volume of this prism by 72 cubic centimeters. Chelsea wants to add eight layers, and Stephen says they only need to add four layers. Their teacher tells them they are both correct. Explain how this is possible.

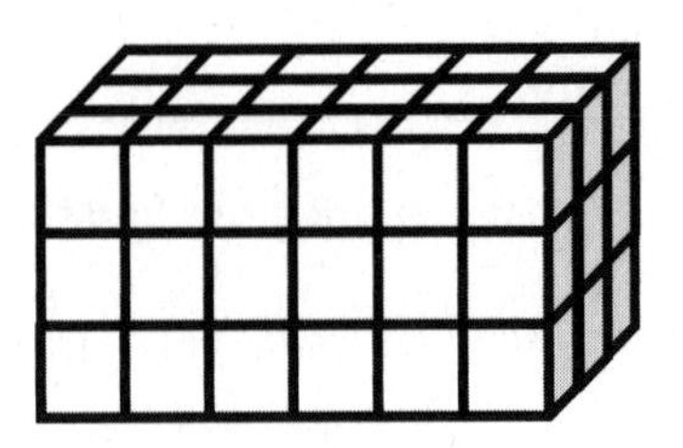

3. Juliana makes a prism 4 inches across and 4 inches wide but only 1 inch tall. She then decides to create layers equal to her first one. Fill in the chart below, and explain how you know the volume of each new prism.

Number of Layers	Volume	Explanation
3		
5		
7		

4. Imagine the rectangular prism below is 4 meters long, 3 meters tall, and 2 meters wide. Draw horizontal lines to show how the prism could be decomposed into layers that are 1 meter in height.

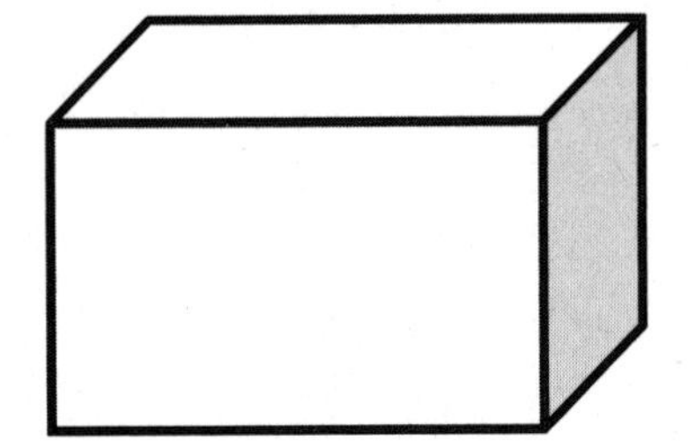

It has ______ layers from top to bottom.

Each horizontal layer contains ________ cubic meters.

The volume of this prism is __________

1. Each rectangular prism is built from centimeter cubes. State the dimensions, and find the volume.

a.

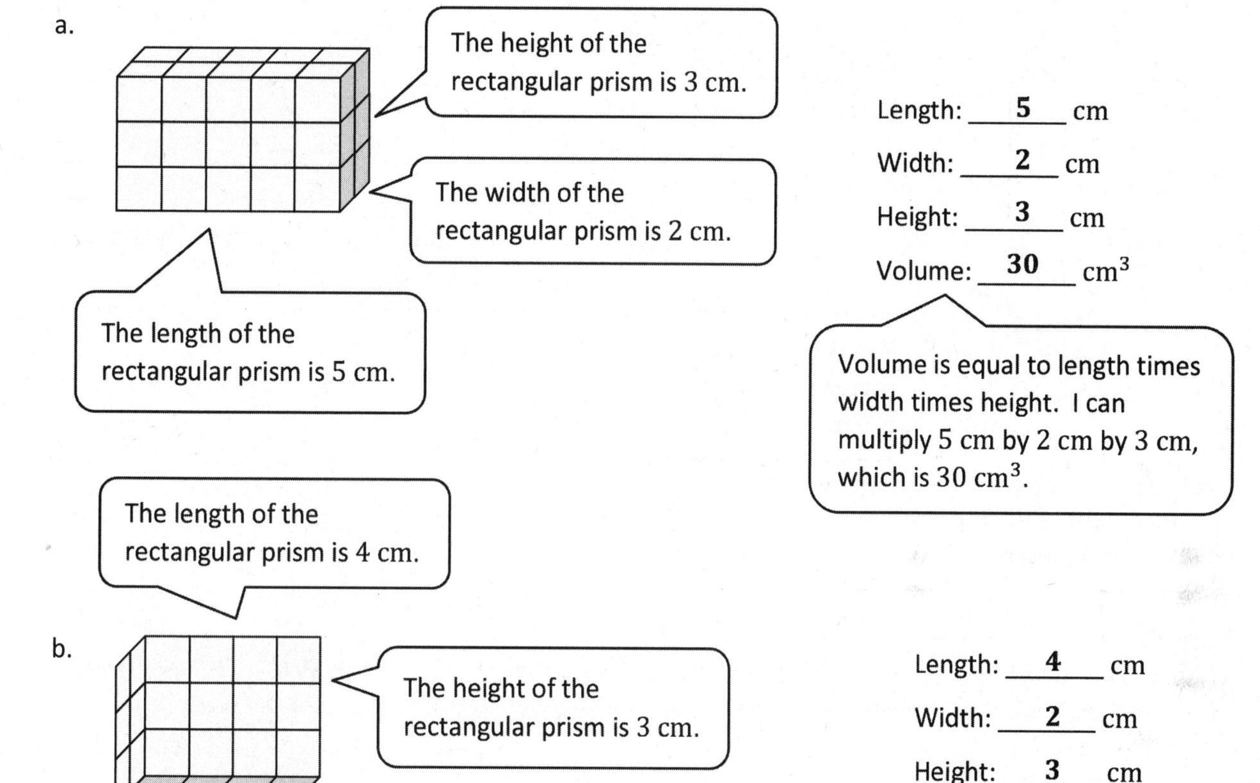

Length: **5** cm

Width: **2** cm

Height: **3** cm

Volume: **30** cm^3

b.

The width of the rectangular prism is 2 cm.

Length: **4** cm

Width: **2** cm

Height: **3** cm

Volume: **24** cm^3

Volume $= l \times w \times h$. I can multiply 4 cm by 2 cm by 3 cm, which is 24 cm^3.

2. Write a multiplication sentence that you could use to calculate the volume for each rectangular prism in Problem 1. Include the units in your sentences.

a. **5 cm × 2 cm × 3 cm = 30 cm^3**

b. **4 cm × 2 cm × 3 cm = 24 cm^3**

3. Calculate the volume of each rectangular prism. Include the units in your number sentences.

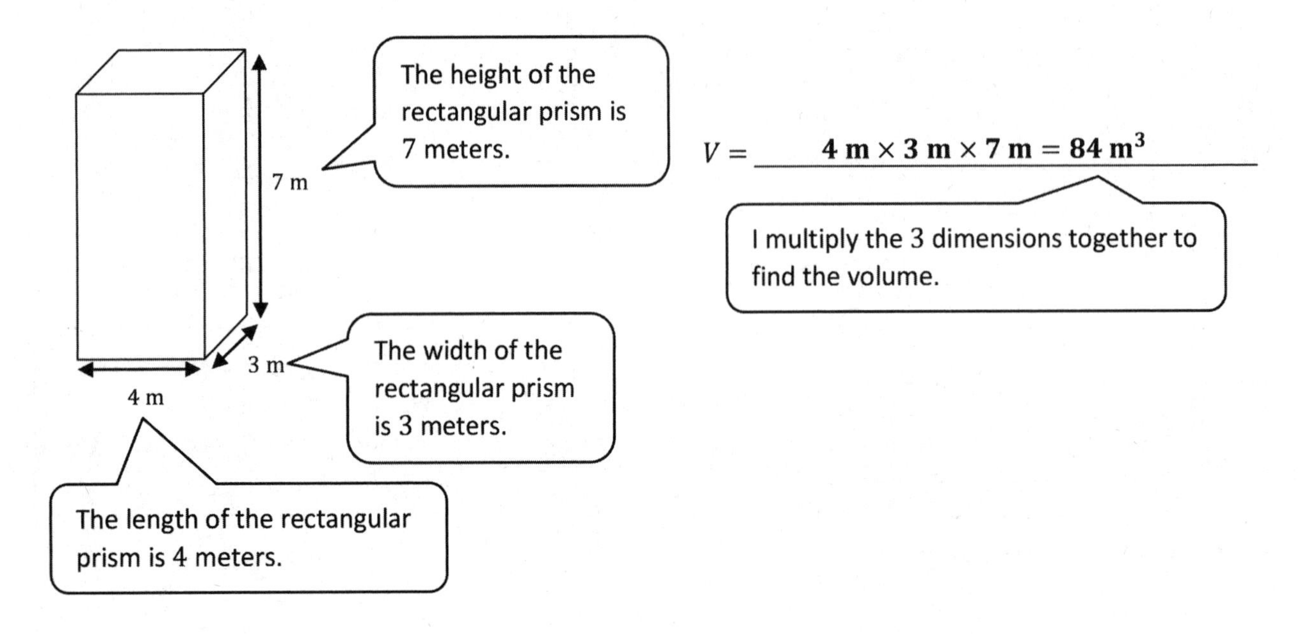

$V =$ $\mathbf{4\ m \times 3\ m \times 7\ m = 84\ m^3}$

4. Meilin is constructing a box in the shape of a rectangular prism to store her small toys. It has a length of 10 inches, a width of 5 inches, and a height of 7 inches. What is the volume of the box?

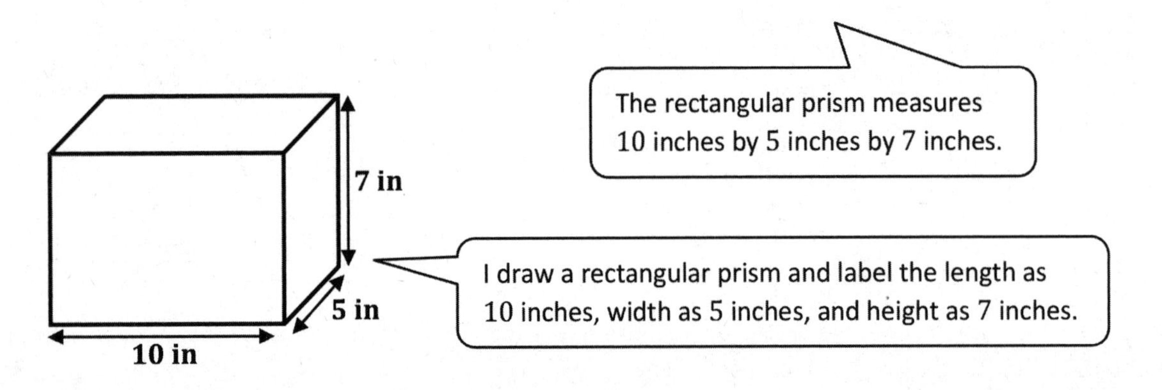

Volume = length × width × height

$\mathbf{V = 10\ in \times 5\ in \times 7\ in = 350\ in^3}$

The volume of the box is 350 cubic inches.

Name ______________________________ Date ______________

1. Each rectangular prism is built from centimeter cubes. State the dimensions, and find the volume.

a.

Length: ______ cm

Width: ______ cm

Height: ______ cm

Volume: ______ cm^3

b.

Length: ______ cm

Width: ______ cm

Height: ______ cm

Volume: ______ cm^3

c.

Length: ______ cm

Width: ______ cm

Height: ______ cm

Volume: ______ cm^3

d.

Length: ______ cm

Width: ______ cm

Height: ______ cm

Volume: ______ cm^3

2. Write a multiplication sentence that you could use to calculate the volume for each rectangular prism in Problem 1. Include the units in your sentences.

a. ______________________ b. ______________________

c. ______________________ d. ______________________

3. Calculate the volume of each rectangular prism. Include the units in your number sentences.

a.

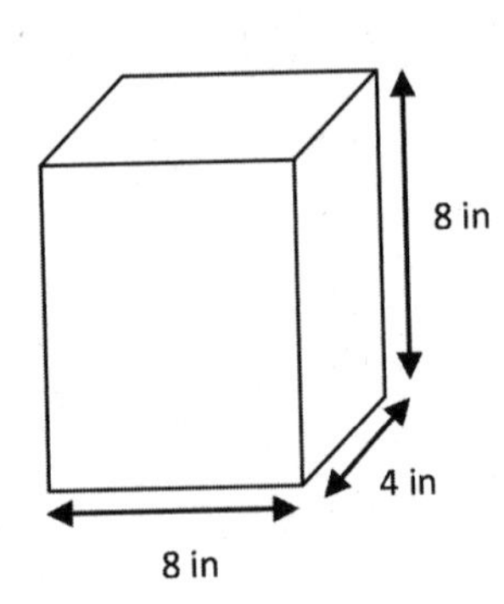

Volume: ______________________________

b.

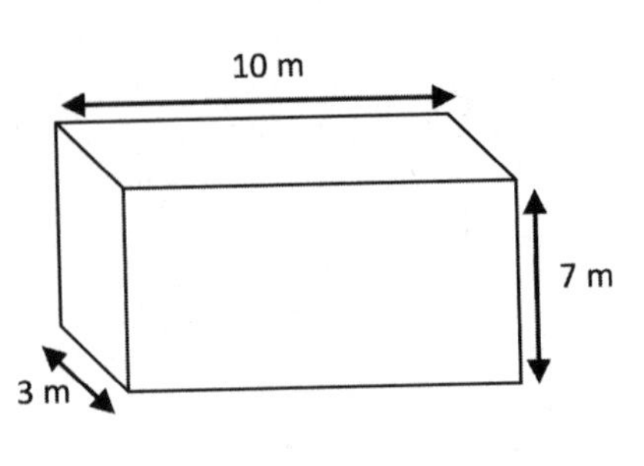

Volume: ______________________________

4. Mrs. Johnson is constructing a box in the shape of a rectangular prism to store clothes for the summer. It has a length of 28 inches, a width of 24 inches, and a height of 30 inches. What is the volume of the box?

5. Calculate the volume of each rectangular prism using the information that is provided.

   a. Face area: 56 square meters

      Height: 4 meters

   b. Face area: 169 square inches

      Height: 14 inches

1. Kevin filled a container with 40 centimeter cubes. Shade the beaker to show how much water the container will hold. Explain how you know.

   It will hold 40 milliliters of water. I know that $1\ \text{cm}^3 = 1\ \text{mL}$. Therefore, $40\ \text{cm}^3$ is equal to $40\ \text{mL}$.

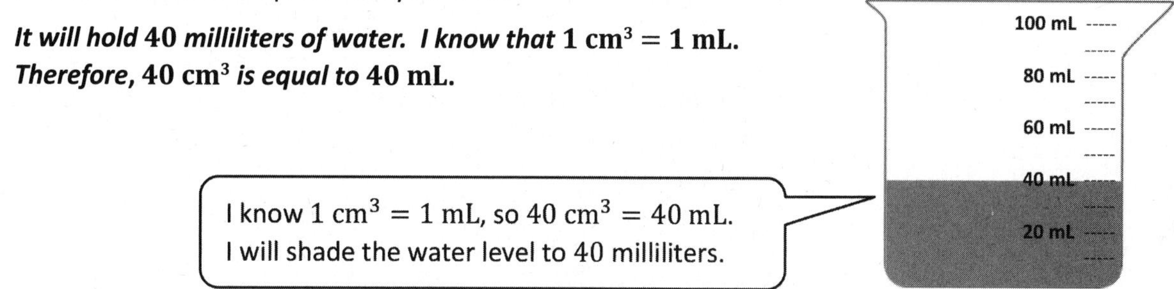

2. A beaker contains $200\ \text{mL}$ of water. Joe wants to pour the water into a container that will hold the water. Which of the containers pictured below could he use? Explain your choices.

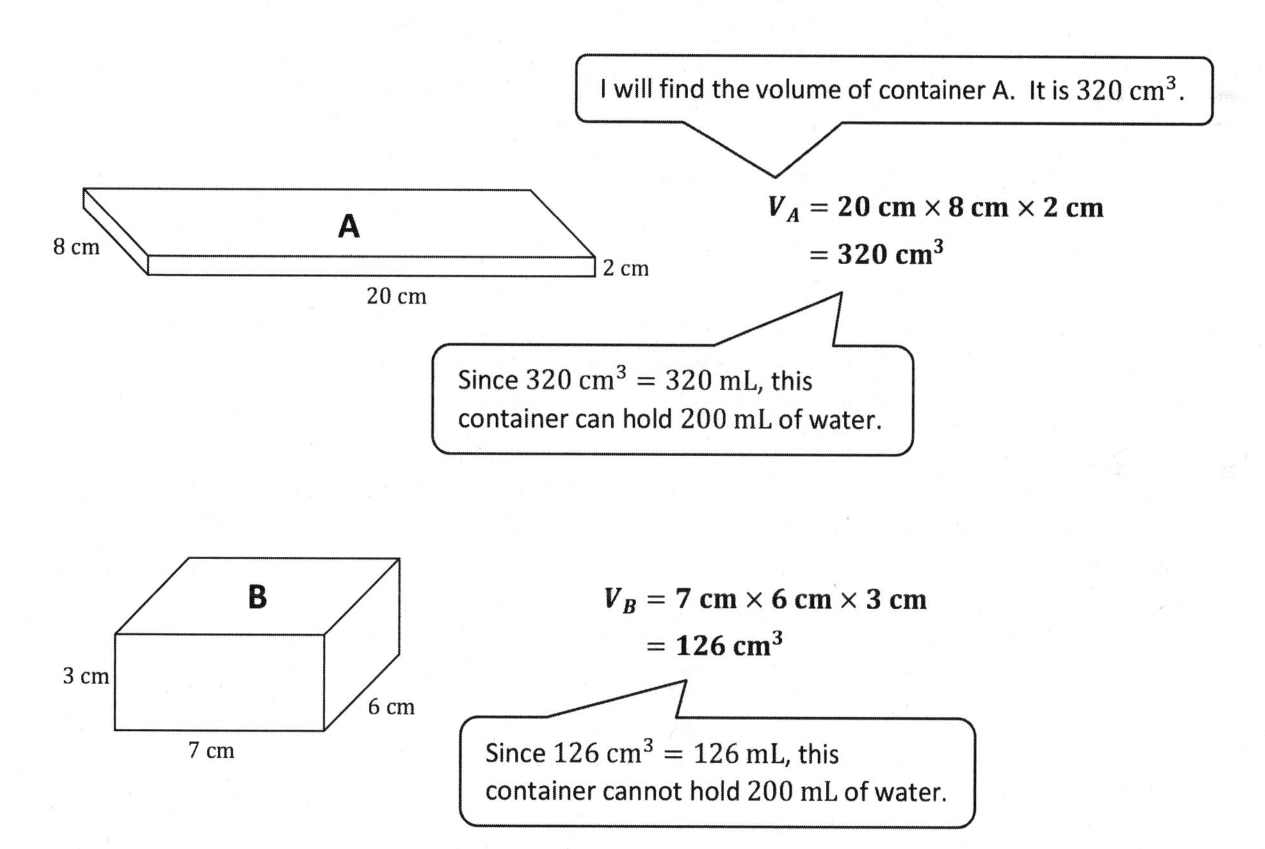

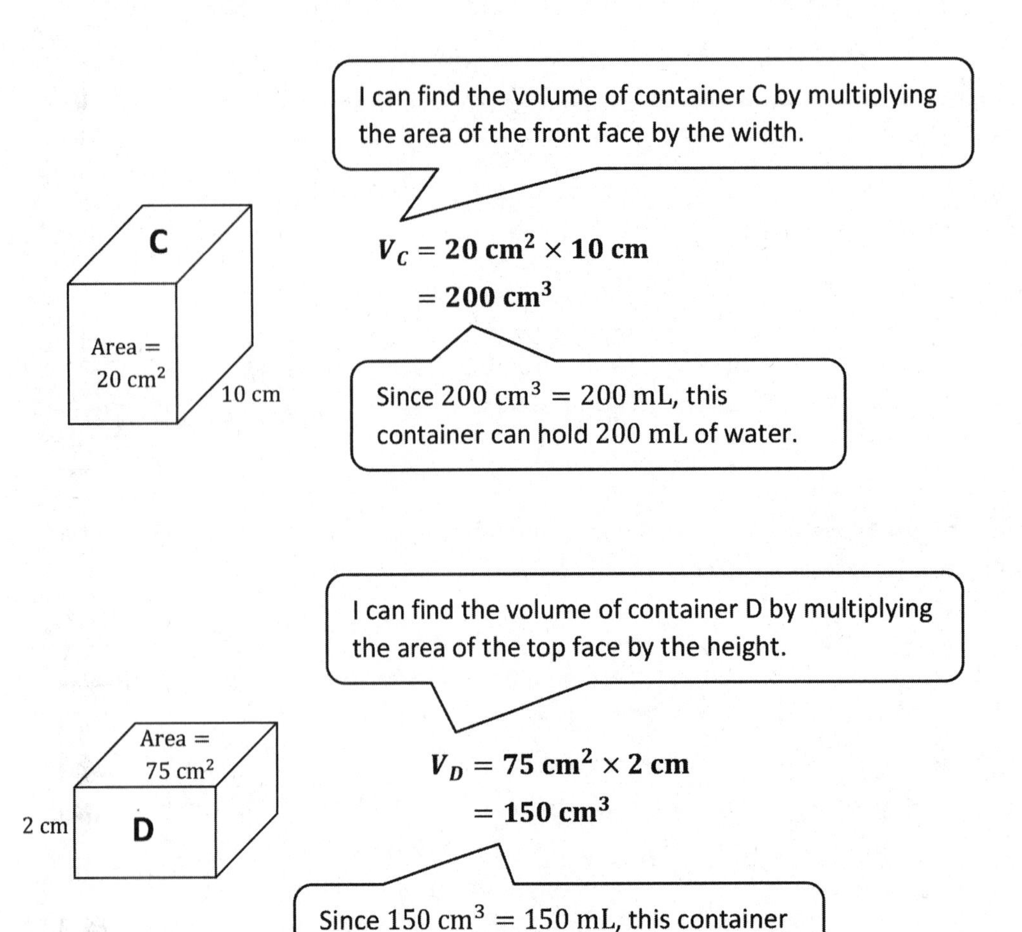

Joe will be able to use container A because the volume is 320 cm^3. He will also be able to use container C because the volume is 200 cm^3. He will not be able to use containers B and D because they are too small.

Name ______________________________ Date ______________

1. Johnny filled a container with 30 centimeter cubes. Shade the beaker to show how much water the container will hold. Explain how you know.

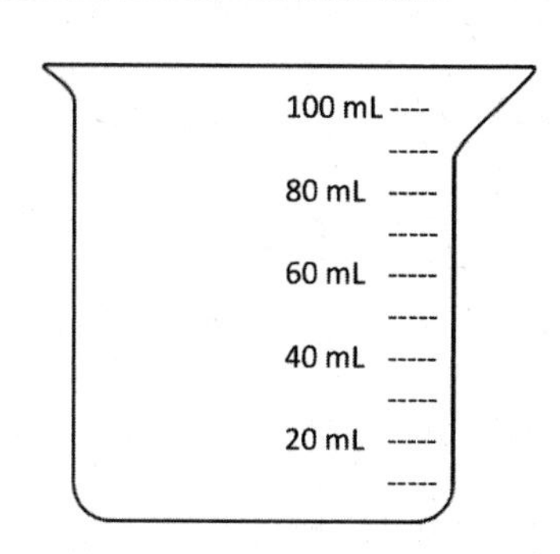

2. A beaker contains 250 mL of water. Jack wants to pour the water into a container that will hold the water. Which of the containers pictured below could he use? Explain your choices.

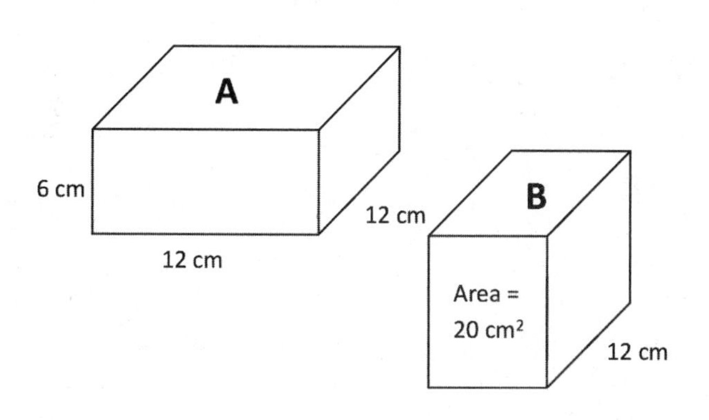

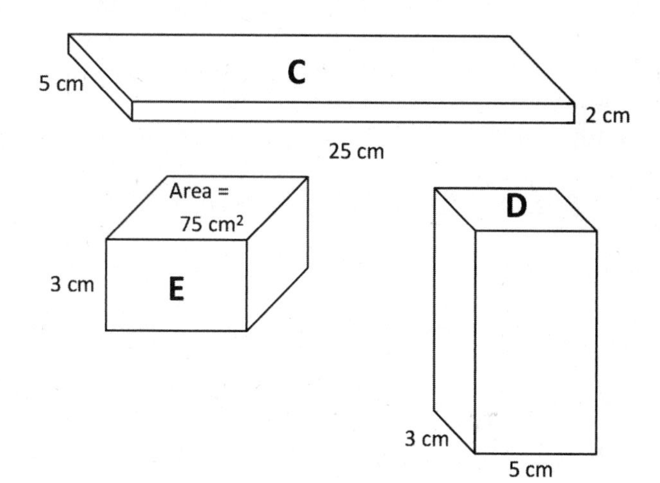

3. On the back of this paper, describe the details of the activities you did in class today. Include what you learned about cubic centimeters and milliliters. Give an example of a problem you solved with an illustration.

1. Find the total volume of the figures, and record your solution strategy.

   a.

The top figure has a length of 5 in and a height of 3 in.

Since the top figure is sitting directly on top of the bottom figure, without any gaps or overlaps, the width of both figures is 4 in.

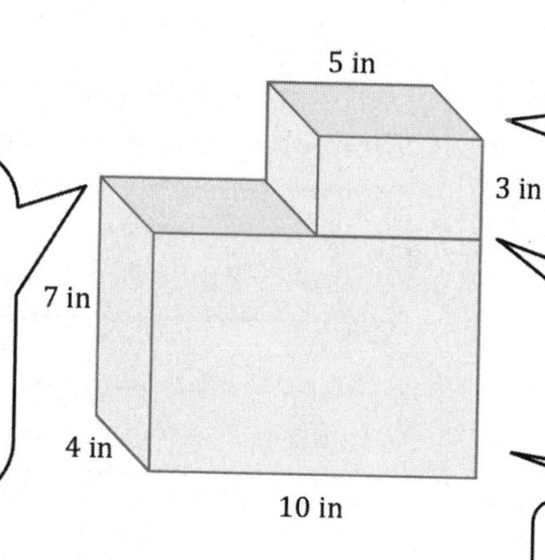

I can find the volume of the top figure. Volume $= 5 \text{ in} \times 4 \text{ in} \times 3 \text{ in} = 60 \text{ in}$

I can find the volume of the bottom figure. Volume $= 10 \text{ in} \times 4 \text{ in} \times 7 \text{ in} = 280 \text{ in}^3$

Volume: **340 in^3**

I will add both figures' volumes together. $60 \text{ in}^3 + 280 \text{ in}^3 = 340 \text{ in}^3$

Solution Strategy:

I found the top figure's volume, 60 in^3, and the bottom figure's volume, 280 in^3. Then, I added both volumes together to get a total of 340 in^3.

b.

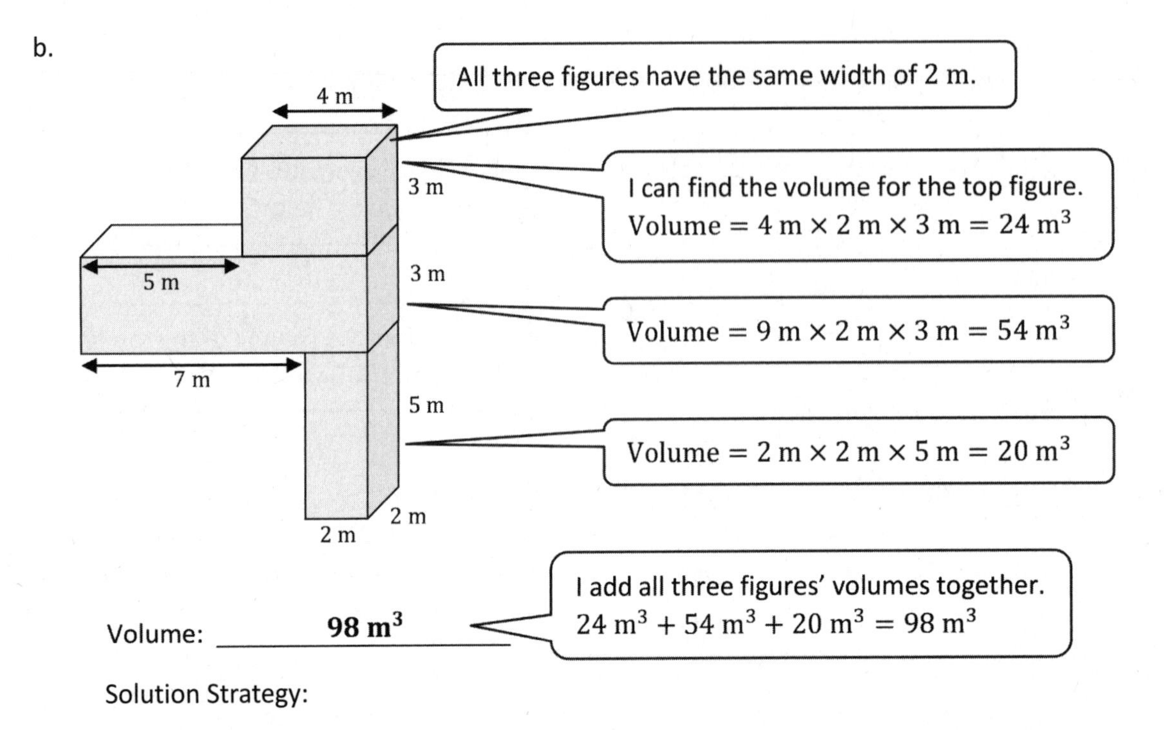

Volume: **98 m^3**

I add all three figures' volumes together.
$24\text{ m}^3 + 54\text{ m}^3 + 20\text{ m}^3 = 98\text{ m}^3$

Solution Strategy:

I found the top figure's volume, 24 m^3, the middle figure's volume, 54 m^3, and the bottom figure's volume, 20 m^3. Then, I added all three volumes together to get a total of 98 m^3.

2. A fish tank has a base area of 65 cm^2 and is filled with water to a depth of 21 cm. If the height of the tank is 30 cm, how much more water will be needed to fill the tank to the brim?

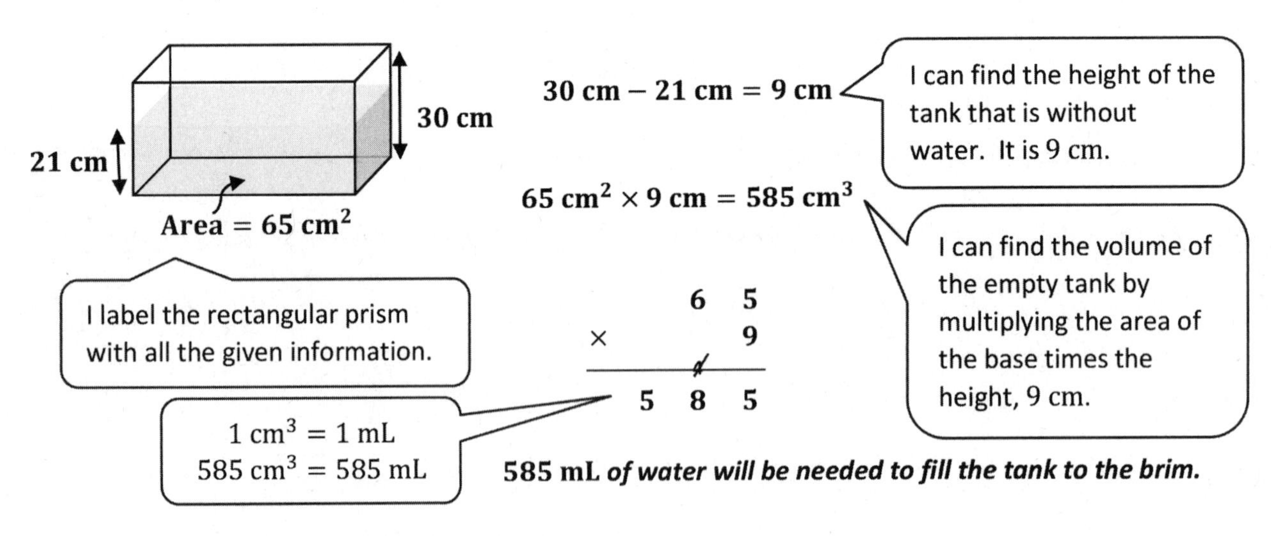

585 mL *of water will be needed to fill the tank to the brim.*

Name ______________________________ Date ______________

1. Find the total volume of the figures, and record your solution strategy.

a.

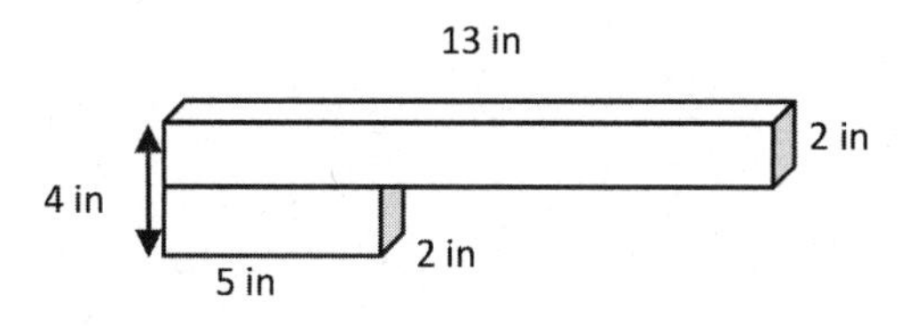

Volume: ______________________

Solution Strategy:

b.

18 cm
2 cm
3 cm
9 cm
7 cm
21 cm

Volume: ______________________

Solution Strategy:

c.

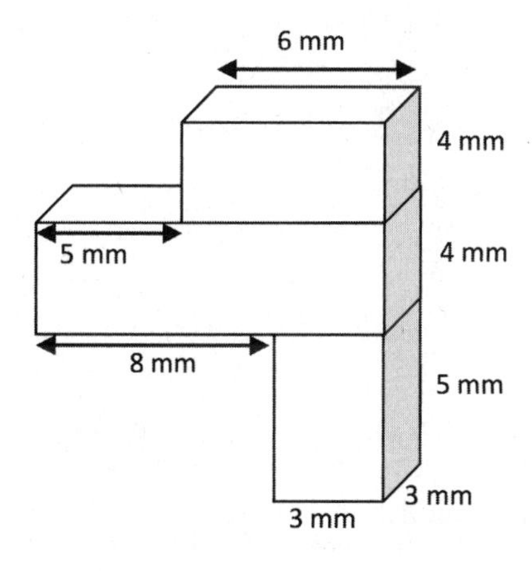

Volume: ______________________

Solution Strategy:

d.

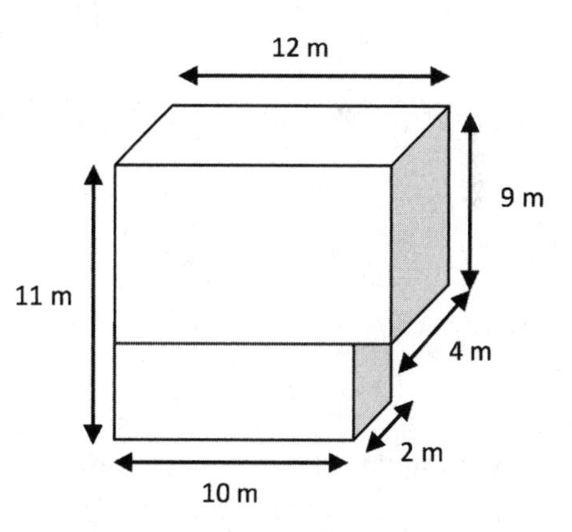

Volume: ______________________

Solution Strategy:

2. The figure below is made of two sizes of rectangular prisms. One type of prism measures 3 inches by 6 inches by 14 inches. The other type measures 15 inches by 5 inches by 10 inches. What is the total volume of this figure?

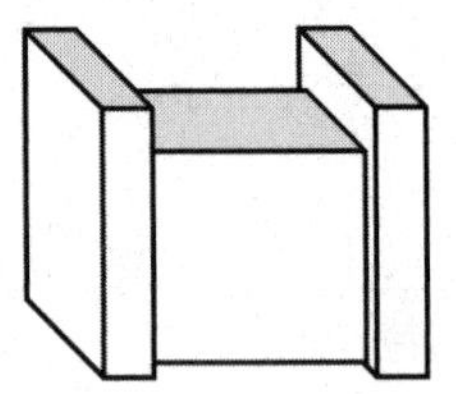

3. The combined volume of two identical cubes is 250 cubic centimeters. What is the measure of one cube's edge?

4. A fish tank has a base area of 45 cm^2 and is filled with water to a depth of 12 cm. If the height of the tank is 25 cm, how much more water will be needed to fill the tank to the brim?

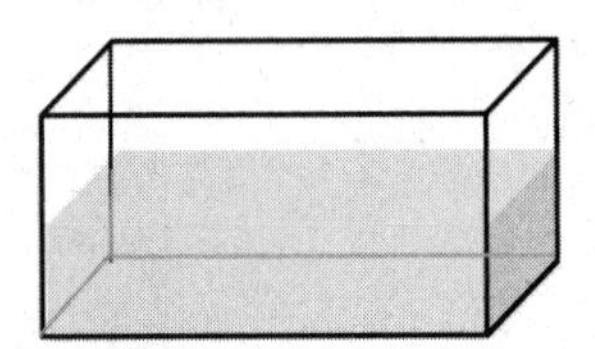

5. Three rectangular prisms have a combined volume of 518 cubic feet. Prism A has one-third the volume of Prism B, and Prisms B and C have equal volume. What is the volume of each prism?

Edwin builds rectangular planters.

1. Edwin's first planter is 6 feet long and 2 feet wide. The container is filled with soil to a height of 3 feet in the planter. What is the volume of soil in the planter? Explain your work using a diagram.

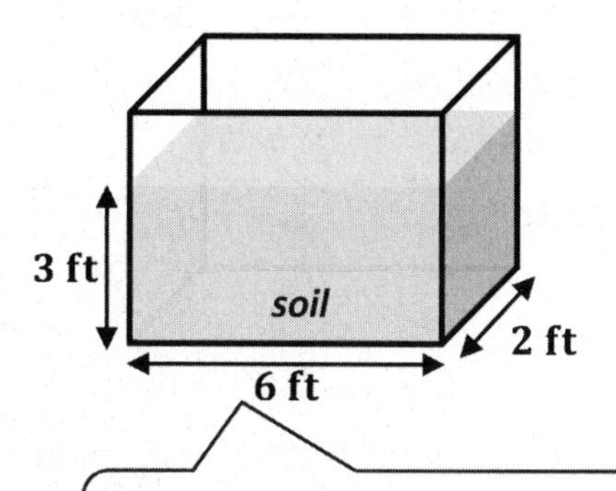

Volume = length × width × height

$V = 6 \text{ ft} \times 2 \text{ ft} \times 3 \text{ ft} = 36 \text{ ft}^3$

The volume of soil in the planter is 36 cubic feet.

I draw a rectangular prism and label all the given information.

I can multiply the length, width, and height of the soil to find the volume of the soil in the planter.

In order to have a volume of 50 cubic feet, I have to think of different factors that I can multiply to get 50. Since volume is three-dimensional, I will have to think of 3 factors.

2. Edwin wants to grow some flowers in two planters. He wants each planter to have a volume of 50 cubic feet, but he wants them to have different dimension. Show two different ways Edwin can make these planters, and draw diagrams with the planters' measurements on them.

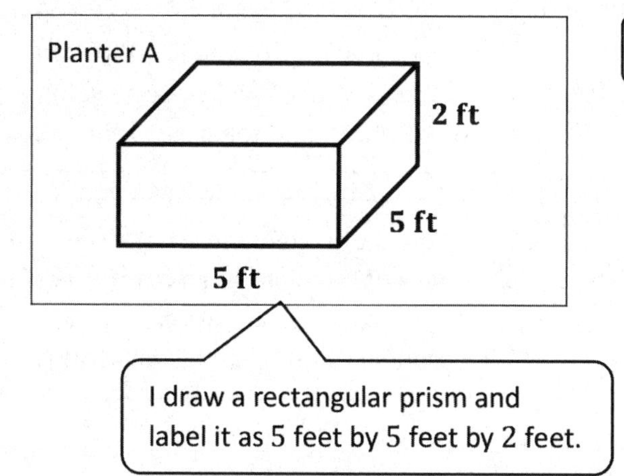

I need to think of 3 factors that give a product of 50.

Volume $= l \times w \times h$

$V = 5 \text{ ft} \times 5 \text{ ft} \times 2 \text{ ft} = 50 \text{ ft}^3$

I draw a rectangular prism and label it as 5 feet by 5 feet by 2 feet.

I can verify my answer by finding the volume for Planter A. The answer is 50 cubic feet.

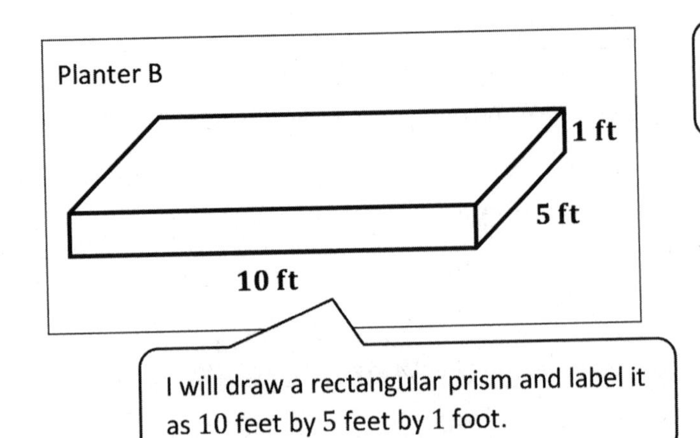

I need the 3 different factors for Planter B.
$10 \times 5 \times 1 = 50$

Volume $= l \times w \times h$

$V = 10 \text{ ft} \times 5 \text{ ft} \times 1 \text{ ft} = 50 \text{ ft}^3$

I will draw a rectangular prism and label it as 10 feet by 5 feet by 1 foot.

In order to have a volume of 30 cubic feet, I have to think of three factors that give a product of 30.

3. Edwin wants to make one planter that extends from the ground to just below his back window. The window starts 3 feet off the ground. If he wants the planer to hold 30 cubic feet of soil, name one way he could build the planter so it is not taller than 3 feet. Explain how you know.

The volume is 30 cubic feet, and one of the dimensions must not be more than 3 feet. So, I will keep the height as 3 feet.

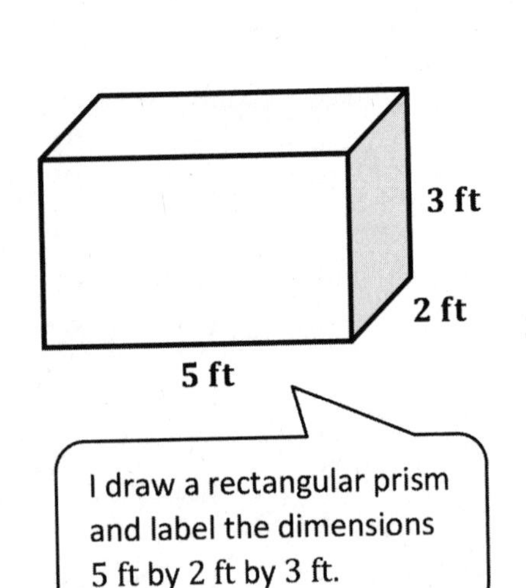

$30 \text{ ft}^3 \div 3 \text{ ft} = 10 \text{ ft}^2$

I already know the volume is 30 ft^3, and the height is 3 ft, so I'll divide the volume by the height to find the area of the base.

I draw a rectangular prism and label the dimensions 5 ft by 2 ft by 3 ft.

$10 \text{ ft}^2 = 5 \text{ ft} \times 2 \text{ ft}$

Length $= 5 \text{ ft}$

Width $= 2 \text{ ft}$

Height $= 3 \text{ ft}$

Now that I know the area of the base of the planter is 10 ft^2, I need to think of two factors that have a product of 10. 5 and 2 will work!

Since Edwin wants to build a planter with a height of 3 ft ***and a volume of*** 30 ft^3***, the base of the planter should have an area of*** 10 ft^2***. I drew a planter with a length of*** 5 ft***, width of*** 2 ft***, and height of*** 3 ft***.***

Name ____ Date _______________________

Wren makes some rectangular display boxes.

1. Wren's first display box is 6 inches long, 9 inches wide, and 4 inches high. What is the volume of the display box? Explain your work using a diagram.

2. Wren wants to put some artwork into three shadow boxes. She knows they all need a volume of 60 cubic inches, but she wants them all to be different. Show three different ways Wren can make these boxes by drawing diagrams and labeling the measurements.

Shadow Box A	Shadow Box B
Shadow Box C	

3. Wren wants to build a box to organize her scrapbook supplies. She has a stencil set that is 12 inches wide that needs to lay flat in the bottom of the box. The supply box must also be no taller than 2 inches. Name one way she could build a supply box with a volume of 72 cubic inches.

4. After all of this organizing, Wren decides she also needs more storage for her soccer equipment. Her current storage box measures 1 foot long by 2 feet wide by 2 feet high. She realizes she needs to replace it with a box with 12 cubic feet of storage, so she doubles the width.

   a. Will she achieve her goal if she does this? Why or why not?

   b. If she wants to keep the height the same, what could the other dimensions be for a 12-cubic-foot storage box?

   c. If she uses the dimensions in part (b), what is the area of the new storage box's floor?

   d. How has the area of the bottom in her new storage box changed? Explain how you know.

1. I have a prism with the dimensions of 8 in by 12 in by 20 in. Calculate the volume of the prism, and then give the dimensions of two different prisms that each have $\frac{1}{4}$ of the volume.

To find $\frac{1}{4}$ of the volume, I can use the original prism's volume divided by 4. $\frac{1}{4}$ of $1{,}920 \text{ in}^3$ is equal to 480 in^3.

	Length	Width	Height	Volume
Original Prism	8 in.	12 in.	20 in.	$\mathbf{1{,}920 \text{ in}^3}$

I multiply the three dimensions to find the original volume.
$8 \text{ in} \times 12 \text{ in} \times 20 \text{ in} = 1{,}920 \text{ in}^3$

	Length	Width	Height	Volume
Prism 1	**2 in.**	**12 in.**	**20 in.**	$\mathbf{480 \text{ in}^3}$

In order to create a volume that is $\frac{1}{4}$ of 1,920, I can change one of the dimensions and keep the others the same.
$\frac{1}{4}$ of $8 \text{ in} = 2 \text{ in}$

$2 \text{ in} \times 12 \text{ in} \times 20 \text{ in} = 480 \text{ in}^3$

	Length	Width	Height	Volume
Prism 2	**8 in.**	**6 in.**	**10 in.**	$\mathbf{480 \text{ in}^3}$

Another way I can create a volume that is $\frac{1}{4}$ of 1,920 is to change two of the dimensions and keep the other the same.
$\frac{1}{2}$ of $12 \text{ in} = 6 \text{ in}$
$\frac{1}{2}$ of $20 \text{ in} = 10 \text{ in}$

Kayla's bedroom has a volume of 800 ft^3.
$10\text{ ft} \times 8\text{ ft} \times 10\text{ ft} = 800\text{ ft}^3$

One way to double the volume is to double one dimension and keep the others the same.

2. Kayla's bedroom has the dimensions of 10 ft by 8 ft by 10 ft. Her den has the same height (10 ft) but double the volume. Give two sets of the possible dimensions of the den and the volume of the den.

Length: $\mathbf{10\text{ ft} \times 2 = 20\text{ ft}}$

I can double the length, $10\text{ ft} \times 2 = 20\text{ ft}$, and keep both the width and the height the same.

Width: **8 ft**

Height: **10 ft**

$\mathbf{Volume = 20\text{ ft} \times 8\text{ ft} \times 10\text{ ft} = 1{,}600\text{ ft}^3}$

1,600 ft^3 is double the original volume of 800 ft^3.

Length: $\mathbf{10\text{ ft} \times 4 = 40\text{ ft}}$

In order to double the volume, I can also quadruple the length and cut the width in half.

Width: $\mathbf{8\text{ ft} \times \frac{1}{2} = 4\text{ ft}}$

Height: **10 ft**

$\mathbf{Volume = 40\text{ ft} \times 4\text{ ft} \times 10\text{ ft} = 1{,}600\text{ ft}^3}$

1,600 ft^3 is double the original volume of 800 ft^3.

Name ______________________________ Date ______________

1. I have a prism with the dimensions of 6 cm by 12 cm by 15 cm. Calculate the volume of the prism, and then give the dimensions of three different prisms that each have $\frac{1}{3}$ of the volume.

	Length	**Width**	**Height**	**Volume**
Original Prism	6 cm	12 cm	15 cm	
Prism 1				
Prism 2				
Prism 3				

2. Sunni's bedroom has the dimensions of 11 ft by 10 ft by 10 ft. Her den has the same height but double the volume. Give two sets of the possible dimensions of the den and the volume of the den.

Find three rectangular prisms around your house. Describe the item you are measuring (e.g., cereal box, tissue box), and then measure each dimension to the nearest whole inch and calculate the volume.

a. Rectangular Prism A

Item: ***Cereal box***

I will measure a cereal box, and then multiply the three dimensions to find the volume.

Height: **12** inches

Length: **8** inches

Width: **3** inches

Volume: **288** cubic inches

$$\begin{aligned}\text{Volume} &= \text{length} \times \text{width} \times \text{height}\\ &= 8 \text{ in} \times 3 \text{ in} \times 12 \text{ in}\\ &= 288 \text{ in}^3\end{aligned}$$

b. Rectangular Prism B

Item: Tissue box

I will measure a tissue box, and then multiply the three dimensions to find the volume.

Height: **3** inches

Length: **9** inches

Width: **5** inches

Volume: **135** cubic inches

$$\begin{aligned}\text{Volume} &= \text{length} \times \text{width} \times \text{height}\\ &= 9 \text{ in} \times 5 \text{ in} \times 3 \text{ in}\\ &= 45 \text{ in}^2 \times 3 \text{ in}\\ &= 135 \text{ in}^3\end{aligned}$$

The volume of the tissue box is 135 cubic inches.

Name ______________________________ Date ____________

1. Find three rectangular prisms around your house. Describe the item you are measuring (cereal box, tissue box, etc.), and then measure each dimension to the nearest whole inch, and calculate the volume.

   a. Rectangular Prism A

      Item:

      Height: ____________ inches

      Length: ____________ inches

      Width: ____________ inches

      Volume: ____________ cubic inches

   b. Rectangular Prism B

      Item:

      Height: ____________ inches

      Length: ____________ inches

      Width: ____________ inches

      Volume: ____________ cubic inches

   c. Rectangular Prism C

      Item:

      Height: ____________ inches

      Length: ____________ inches

      Width: ____________ inches

      Volume: ____________ cubic inches

1. Alex tiled some rectangles using square units. Sketch the rectangles if necessary. Fill in the missing information, and then confirm the area by multiplying.

Rectangle A:

I look at Rectangle A's dimensions, 4 units by $2\frac{1}{2}$ units.

Rectangle A is

4 units long by $2\frac{1}{2}$ unit wide.

Area = **10** square units

I can draw a length of 4 units.

I can draw a rectangle and show a width of $2\frac{1}{2}$ units.

4 units

2 units

$\frac{1}{2}$ ***unit***

I can count the halves and see that there are 4 half square units, which is the same as 2 square units. I can multiply too.

4 units $\times \frac{1}{2}$ unit $=$ 2 square units

I can count the squares and see that there are 8 whole square units. I can multiply too.

4 units $\times$ 2 units $=$ 8 square units

8 square units + 2 square units = 10 square units

4 *units* $\times$ $\mathbf{2\frac{1}{2}}$ ***units***

I can confirm the area by multiplying the length and width.

The area of Rectangle A is 10 square units.

$$(4 \times 2) + \left(4 \times \frac{1}{2}\right)$$

$$= 8 + \frac{4}{2}$$

$$= 8 + 2$$

$$= 10$$

I can use the rectangle I drew and the distributive property to help me multiply.

4 units $\times$ 2 units $=$ 8 square units

4 units $\times \frac{1}{2}$ unit $= \frac{4}{2}$ square units $=$ 2 square units

2. Juanita made a mosaic from different colored rectangular tiles. Two blue tiles measured $2\frac{1}{2}$ inches × 3 inches. Five white tiles measured 3 inches × $2\frac{1}{4}$ inches. What is the area of the whole mosaic in square inches?

I can find the area of one blue tile.

$2\frac{1}{2}\text{ in} \times 3\text{ in}$

$(2 \times 3) + \left(\frac{1}{2} \times 3\right)$

$= 6 + \frac{3}{2}$

$= 6 + 1\frac{1}{2}$

$= 7\frac{1}{2}$

The area of 1 blue tile is $7\frac{1}{2}\text{ in}^2$.

To find the area of the two blue tiles, I can multiply the area by 2.

$1\ unit = 7\frac{1}{2}\text{ in}^2$

$2\ units = 2 \times 7\frac{1}{2}\text{ in}^2$

$= (2 \times 7) + \left(2 \times \frac{1}{2}\right)$

$= 14 + \frac{2}{2}$

$= 14 + 1$

$= 15$

The area of 2 blue tile is 15 in^2.

I can find the area of one white tile.

$3\text{ in} \times 2\frac{1}{4}\text{ in}$

$(3 \times 2) + \left(3 \times \frac{1}{4}\right)$

$= 6 + \frac{3}{4}$

$= 6\frac{3}{4}$

The area of 1 white tile is $6\frac{3}{4}\text{ in}^2$.

To find the area of five white tiles, I can multiply the area by 5.

$1\ unit = 6\frac{3}{4}\text{ in}^2$

$5\ units = 5 \times 6\frac{3}{4}\text{ in}^2$

$= (5 \times 6) + \left(5 \times \frac{3}{4}\right)$

$= 30 + \frac{15}{4}$

$= 30 + 3\frac{3}{4}$

$= 33\frac{3}{4}$

The area of 5 white tiles is $33\frac{3}{4}\text{ in}^2$.

$33\frac{3}{4}\text{ in}^2 + 15\text{ in}^2 = 48\frac{3}{4}\text{ in}^2$

I can add the two areas together to find the area of the entire mosaic.

The area of the whole mosaic is $48\frac{3}{4}$ *square inches.*

Name ______________________________ Date ________________

1. John tiled some rectangles using square units. Sketch the rectangles if necessary. Fill in the missing information, and then confirm the area by multiplying.

a. **Rectangle A:**

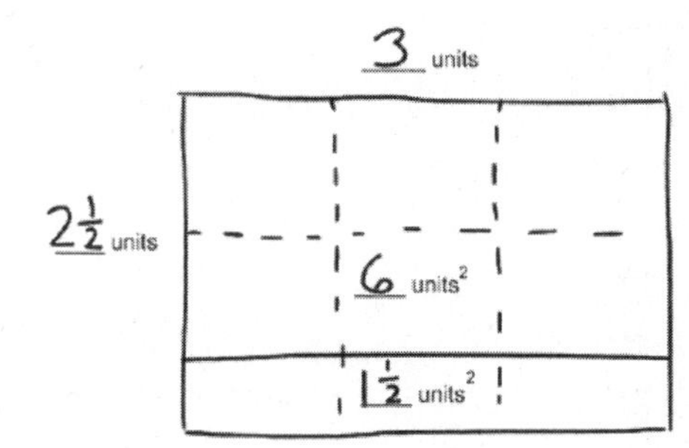

Rectangle A is

3 units long 2 1/2 units wide

Area = ________ units²

b. **Rectangle B:**

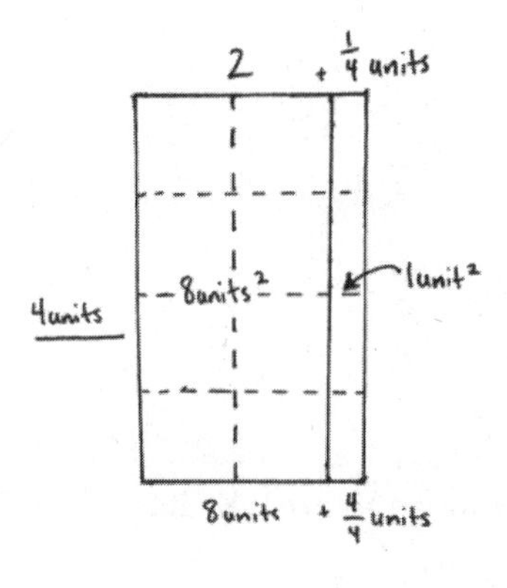

Rectangle B is

________ units long ________ units wide

Area = ________ units²

c. **Rectangle C:**

Rectangle C is

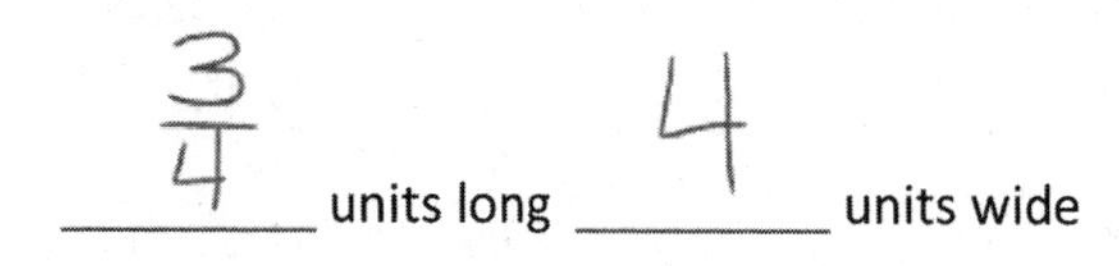

________ units long ________ units wide

Area = ________ units²

d. **Rectangle D:**

Rectangle D is

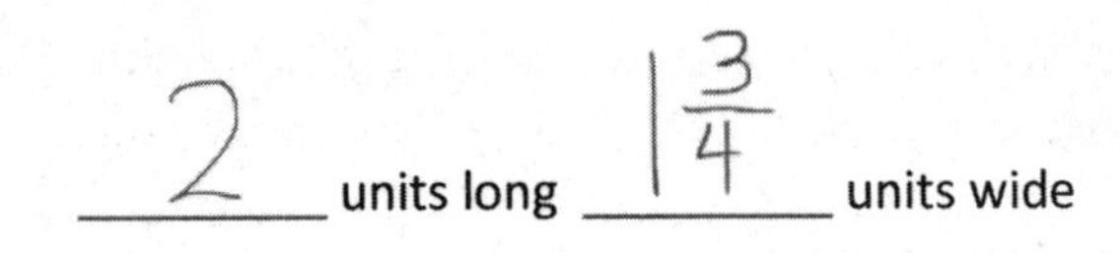

_______ units long _______ units wide

Area = _______ units^2

2. Rachel made a mosaic from different color rectangular tiles. Three tiles measured $3\frac{1}{2}$ inches × 3 inches. Six tiles measured 4 inches × $3\frac{1}{4}$ inches. What is the area of the whole mosaic in square inches?

3. A garden box has a perimeter of $27\frac{1}{2}$ feet. If the length is 9 feet, what is the area of the garden box?

1. Cindy tiled the following rectangles using square units. Sketch the rectangles, and find the areas. Then, confirm the area by multiplying.

   a. **Rectangle A:**

I look at Rectangle A's dimensions, $3\frac{1}{2}$ units by $2\frac{1}{2}$ units.

I can draw a length of $3\frac{1}{2}$ units.

Rectangle A is $3\frac{1}{2}$ units long by $2\frac{1}{2}$ units wide.

Area = **$8\frac{3}{4}$** units^2

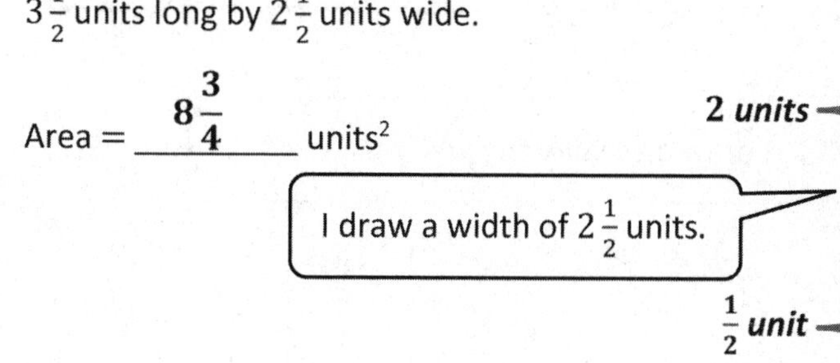

I draw a width of $2\frac{1}{2}$ units.

3 *units* **$\frac{1}{2}$ *unit*** **2 *units*** **$\frac{1}{2}$ *unit***

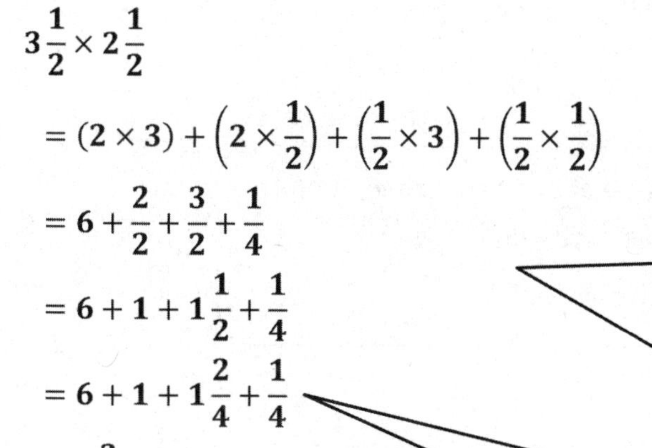

$$3\frac{1}{2} \times 2\frac{1}{2}$$

$$= (2 \times 3) + \left(2 \times \frac{1}{2}\right) + \left(\frac{1}{2} \times 3\right) + \left(\frac{1}{2} \times \frac{1}{2}\right)$$

$$= 6 + \frac{2}{2} + \frac{3}{2} + \frac{1}{4}$$

$$= 6 + 1 + 1\frac{1}{2} + \frac{1}{4}$$

$$= 6 + 1 + 1\frac{2}{4} + \frac{1}{4}$$

$$= 8\frac{3}{4}$$

I can look at the rectangle above to help me multiply.

2 units × 3 units = 6 units^2

2 units × $\frac{1}{2}$ unit = $\frac{2}{2}$ unit^2 = 1 unit^2

$\frac{1}{2}$ unit × 3 units = $\frac{3}{2}$ units^2 = $1\frac{1}{2}$ units^2

$\frac{1}{2}$ unit × $\frac{1}{2}$ unit = $\frac{1}{4}$ unit^2

I rename $1\frac{1}{2}$ as $1\frac{2}{4}$ so I can add.

The area of Rectangle A is $8\frac{3}{4}$ square units.

b. **Rectangle B:**

Rectangle B is

$3\frac{1}{3}$ units long by $\frac{3}{4}$ unit wide.

Area = $2\frac{1}{2}$ units²

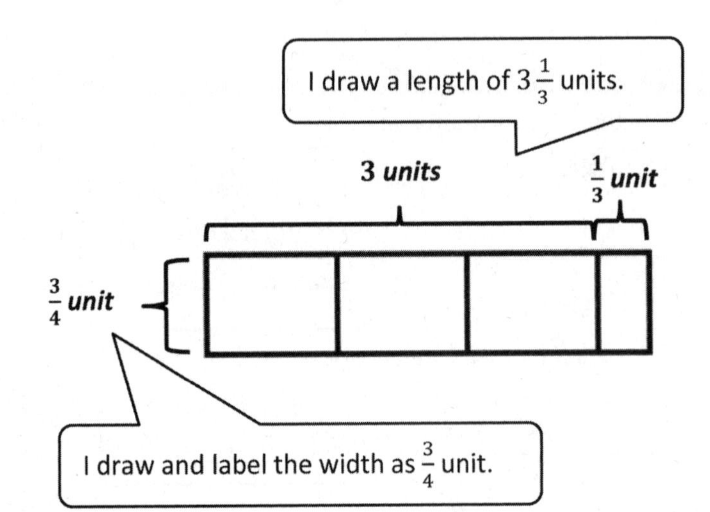

I can multiply to find the area.

$$3\frac{1}{3} \times \frac{3}{4}$$
$$= \left(\frac{3}{4} \times 3\right) + \left(\frac{3}{4} \times \frac{1}{3}\right)$$
$$= \frac{9}{4} + \frac{3}{12}$$
$$= 2\frac{1}{4} + \frac{1}{4}$$
$$= 2\frac{2}{4}$$
$$= 2\frac{1}{2}$$

I can look at the rectangle above to help me multiply.
$\frac{3}{4}$ unit × 3 units = $\frac{9}{4}$ unit^2 = $2\frac{1}{4}$ unit^2
$\frac{3}{4}$ unit × $\frac{1}{3}$ unit = $\frac{3}{12}$ unit^2 = $\frac{1}{4}$ unit^2

The area of Rectangle B is $2\frac{1}{2}$ square units.

2. A Square has a perimeter of 36 inches. What is the area of the square?

All four sides are equal in a square.

Since the perimeter of the square is 36 inches, I will use 36 inches divided by 4 to find the length of one side. $36 \text{ inches} \div 4 = 9 \text{ inches}$

?

Area = ?

$\textbf{Perimeter} = \textbf{36 in}$

$\textbf{36 in} \div \textbf{4} = \textbf{9 in}$

Area is equal to length times width. I will multiply 9 inches times 9 inches to find an area of 81 square inches.

$\textbf{Area} = \textbf{length} \times \textbf{width}$

$= \textbf{9 in} \times \textbf{9 in}$

$= \textbf{81 in}^2$

I can draw a square and label both the area and the side length with a question mark.

The area of the square is $\textbf{81 in}^2$**.**

Name ____________________ Date ____________

1. Kristen tiled the following rectangles using square units. Sketch the rectangles, and find the areas. Then, confirm the area by multiplying. Rectangle A has been sketched for you.

a. **Rectangle A:**

	2 units	$\frac{3}{4}$ unit
1 unit	2 units^2	$\frac{3}{4}$ units^2
$\frac{1}{2}$ unit	1 units^2	$\frac{3}{8}$ units^2

Rectangle A is

________ units long × ________ units wide

Area = ________ units^2

b. **Rectangle B:**

Rectangle B is

$2\frac{1}{2}$ units long × $\frac{3}{4}$ unit wide

Area = ________ units^2

c. **Rectangle C:**

Rectangle C is

$3\frac{1}{3}$ units long × $2\frac{1}{2}$ units wide

Area = ________ units^2

d. **Rectangle D:**

Rectangle D is

$3\frac{1}{2}$ units long × $2\frac{1}{4}$ units wide

Area = __________ units²

2. A square has a perimeter of 25 inches. What is the area of the square?

1. Measure the rectangle to the nearest $\frac{1}{4}$ inch with your ruler, and label the dimensions. Use the area model to find the area.

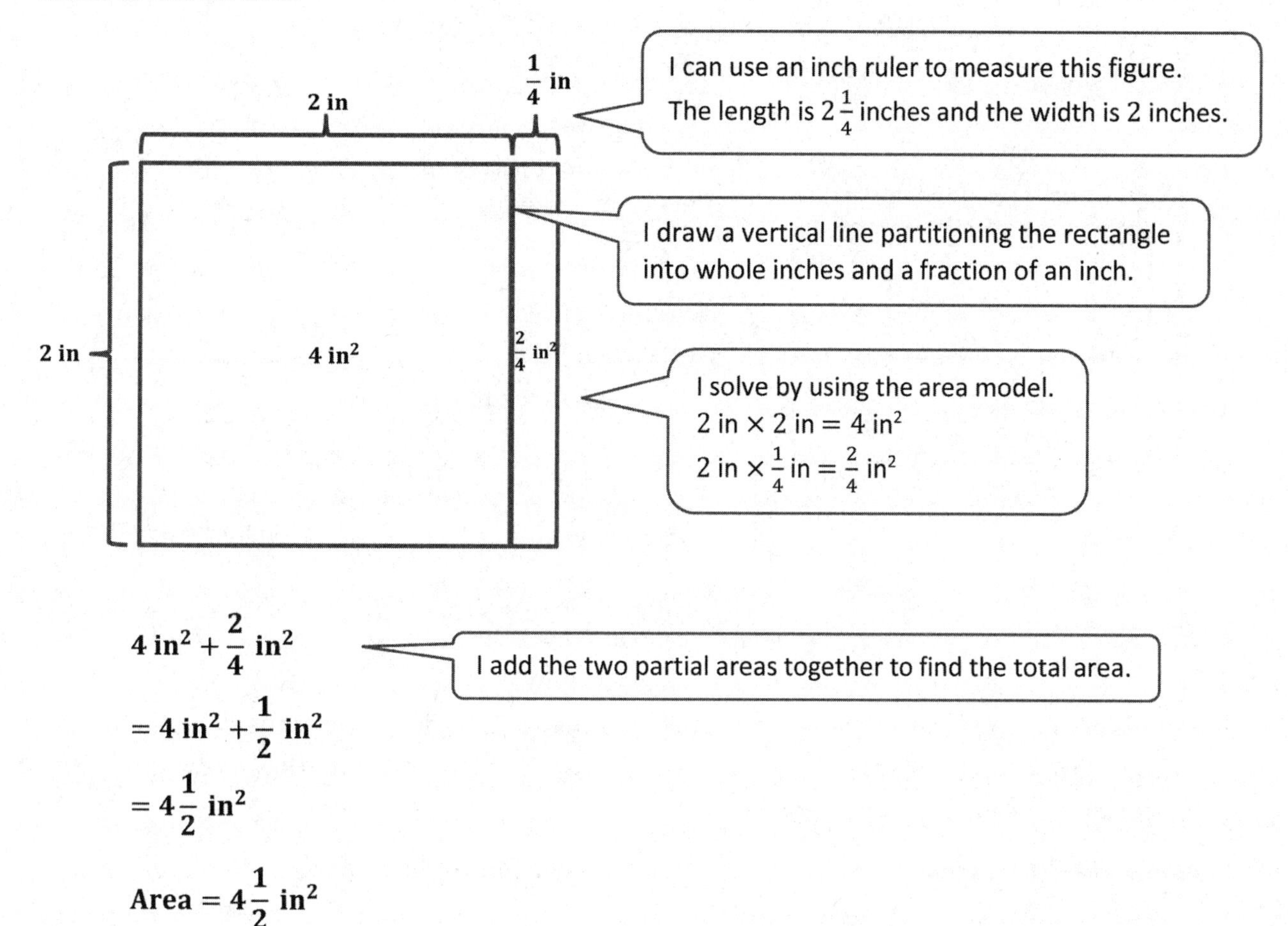

$$4 \text{ in}^2 + \frac{2}{4} \text{ in}^2$$

I add the two partial areas together to find the total area.

$$= 4 \text{ in}^2 + \frac{1}{2} \text{ in}^2$$

$$= 4\frac{1}{2} \text{ in}^2$$

$$\text{Area} = 4\frac{1}{2} \text{ in}^2$$

2. Find the area of rectangle with the following dimensions. Explain your thinking using the area model.

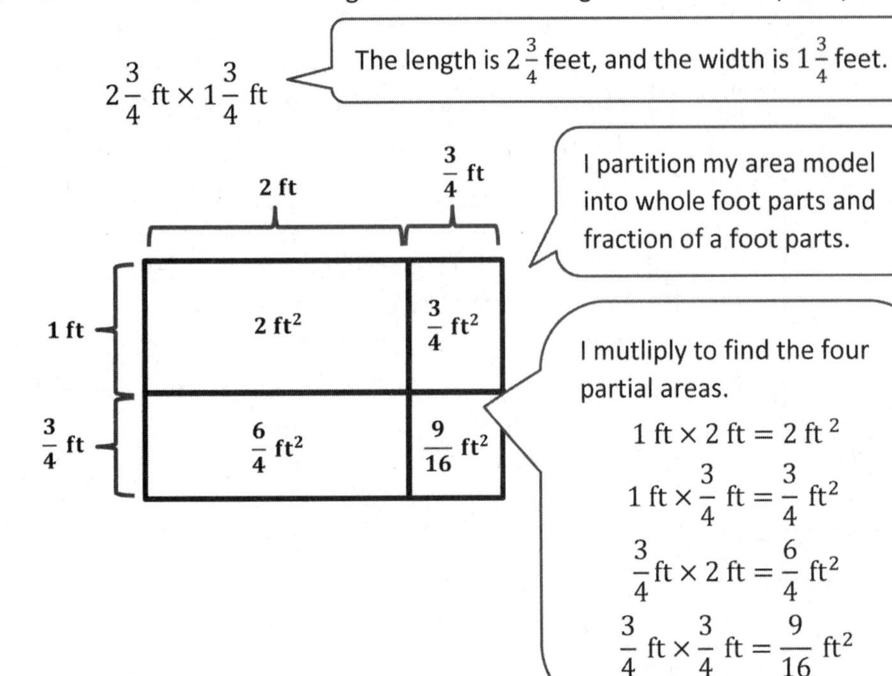

$$\mathbf{2 + \frac{3}{4} + \frac{6}{4} + \frac{9}{16}}$$

$$\mathbf{= 2 + \frac{9}{4} + \frac{9}{16}}$$

$$\mathbf{= 2 + 2\frac{1}{4} + \frac{9}{16}}$$

$$\mathbf{= 2 + 2\frac{4}{16} + \frac{9}{16}}$$

$$\mathbf{= 4\frac{13}{16}}$$

$$\mathbf{Area = 4\frac{13}{16}\ ft^2}$$

3. Zikera is putting carpet in her house. She wants to carpet her living room, which measures $12 \text{ ft} \times 10\frac{1}{2} \text{ ft}$. She also wants to carpet her bedroom, which is $10 \text{ ft} \times 7\frac{1}{2} \text{ ft}$. How many square feet of carpet will she need to cover both rooms?

Area of the living room:

$$\mathbf{12\ ft \times 10\frac{1}{2}\ ft}$$

$$\mathbf{(12 \times 10) + \left(12 \times \frac{1}{2}\right)}$$

$$\mathbf{= 120 + 6}$$

$$\mathbf{= 126}$$

$$\mathbf{Area = 126\ ft^2}$$

I find the area of the living room by multiplying the length and width. It is 126 square feet.

Area of the bedroom:

$$\mathbf{10\ ft \times 7\frac{1}{2}\ ft}$$

$$\mathbf{10 \times \frac{15}{2}}$$

$$\mathbf{= \frac{150}{2}}$$

$$\mathbf{= 75}$$

$$\mathbf{Area = 75\ ft^2}$$

I find the area of the bedroom by multiplying the length and width. It is 75 square feet.

$$\mathbf{126\ ft^2 + 75\ ft^2 = 201\ ft^2}$$

She will need* 201 *square feet of carpet to cover both rooms.

I combine both the area of both rooms to find the total area. The total is 201 square feet.

Name ______________________________ Date ______________

1. Measure each rectangle to the nearest $\frac{1}{4}$ inch with your ruler, and label the dimensions. Use the area model to find the area.

a.

b.

c.

d.

e.

2. Find the area of rectangles with the following dimensions. Explain your thinking using the area model.

a. $2\frac{1}{4}$ yd × $\frac{1}{4}$ yd

b. $2\frac{1}{2}$ ft × $1\frac{1}{4}$ ft

3. Kelly buys a tarp to cover the area under her tent. The tent is 4 feet wide and has an area of 31 square feet. The tarp she bought is $5\frac{1}{3}$ feet by $5\frac{3}{4}$ feet. Can the tarp cover the area under Kelly's tent? Draw a model to show your thinking.

4. Shannon and Leslie want to carpet a $16\frac{1}{2}$-ft by $16\frac{1}{2}$-ft square room. They cannot put carpet under an entertainment system that juts out. (See the drawing below.)

a. In square feet, what is the area of the space with no carpet?

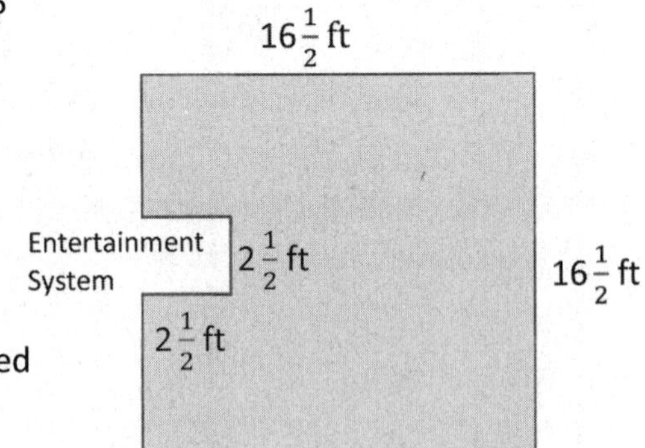

b. How many square feet of carpet will Shannon and Leslie need to buy?

1. Find the area of the following rectangles. Draw an area model if it helps you.

a. $\frac{35}{4}$ ft $\times 2\frac{3}{7}$ ft

I can use multiplication to find the area.

$\frac{35}{4} \times \frac{17}{7}$

I can rename $2\frac{3}{7}$ as a fraction greater than one, $\frac{17}{7}$.

$= \frac{{}^{5}\cancel{35} \times 17}{4 \times \cancel{7}^{1}}$

$= \frac{5 \times 17}{4 \times 1}$

35 and 7 have a common factor of 7. $35 \div 7 = 5$, and $7 \div 7 = 1$. The new numerator is 5×17, and the denominator is 4×1.

$= \frac{85}{4}$

$= 21\frac{1}{4}$

I can use division to convert from a fraction to a mixed number. 85 divided by 4 is equal to $21\frac{1}{4}$.

Area $= 21\frac{1}{4}\text{ ft}^2$

b. $4\frac{2}{3}$ m $\times 2\frac{3}{5}$ m

I use the area model to solve this problem.

	4 m	$\frac{2}{3}$ m
2 m	8 m^2	$\frac{4}{3}\text{ m}^2 = 1\frac{1}{3}\text{ m}^2$
$\frac{3}{5}$ m	$\frac{12}{5}\text{ m}^2 = 2\frac{2}{5}\text{ m}^2$	$\frac{6}{15}\text{ m}^2$

I can multiply to find all four partial products.
$2\text{ m} \times 4\text{ m} = 8\text{ m}^2$
$2\text{ m} \times \frac{2}{3}\text{ m} = \frac{4}{3}\text{ m}^2 = 1\frac{1}{3}\text{ m}^2$
$\frac{3}{5}\text{ m} \times 4\text{ m} = \frac{12}{5}\text{ m}^2 = 2\frac{2}{5}\text{ m}^2$
$\frac{3}{5}\text{ m} \times \frac{2}{3}\text{ m} = \frac{6}{15}\text{ m}^2$

I can add all four partial products to find the area.

$8\text{ m}^2 + 1\frac{1}{3}\text{ m}^2 + 2\frac{2}{5}\text{ m}^2 + \frac{6}{15}\text{ m}^2$

$= 11\text{ m}^2 + \frac{1}{3}\text{ m}^2 + \frac{2}{5}\text{ m}^2 + \frac{6}{15}\text{ m}^2$

$= 11\text{ m}^2 + \frac{5}{15}\text{ m}^2 + \frac{6}{15}\text{ m}^2 + \frac{6}{15}\text{ m}^2$

$= 11\text{ m}^2 + \frac{17}{15}\text{ m}^2$

$= 11\text{ m}^2 + 1\frac{2}{15}\text{ m}^2$

$= 12\frac{2}{15}\text{ m}^2$

Area $= 12\frac{2}{15}\text{ m}^2$

2. Meigan is cutting rectangles out of fabric to make a quilt. If the rectangles are $4\frac{3}{4}$ inches long and $2\frac{1}{2}$ inches wide, what is the area of five such rectangles?

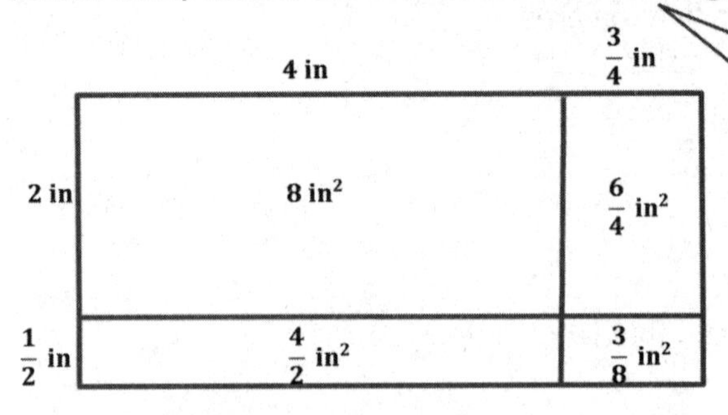

I can find the area of 1 rectangle, and then multiply by 5 to find the total area of 5 rectangles.

I draw an area model to help solve for the area of 1 rectangle.

$$4\frac{3}{4} \times 2\frac{1}{2}$$

$$= (4 \times 2) + \left(4 \times \frac{1}{2}\right) + \left(\frac{3}{4} \times 2\right) + \left(\frac{3}{4} \times \frac{1}{2}\right)$$

$$= 8 + \frac{4}{2} + \frac{6}{4} + \frac{3}{8}$$

$$= 8 + 2 + 1\frac{2}{4} + \frac{3}{8}$$

$$= 11 + \frac{4}{8} + \frac{3}{8}$$

$$= 11\frac{7}{8}$$

I can add up the four partial products. The area of 1 rectangle is $11\frac{7}{8}$ square inches.

$1\ \textit{unit} = 11\frac{7}{8}\ \text{in}^2$

$5\ \textit{units} = 5 \times 11\frac{7}{8}\ \text{in}^2$

The area of 1 rectangle or 1 unit is equal to $11\frac{7}{8}$ square inches. I can multiply by 5 to find the area of 5 rectangles or 5 units.

$$(5 \times 11) + \left(5 \times \frac{7}{8}\right)$$

$$= 55 + \frac{35}{8}$$

$$= 55 + 4\frac{3}{8}$$

$$= 59\frac{3}{8}$$

The area of five rectangles is $59\frac{3}{8}$ square inches.

Name ______________________________ Date ______________

1. Find the area of the following rectangles. Draw an area model if it helps you.

a. $\frac{8}{3}$ cm × $\frac{24}{4}$ cm

b. $\frac{32}{5}$ ft × $3\frac{3}{8}$ ft

c. $5\frac{4}{6}$ in × $4\frac{3}{5}$ in

d. $\frac{5}{7}$ m × $6\frac{3}{5}$ m

2. Chris is making a tabletop from some leftover tiles. He has 9 tiles that measure $3\frac{1}{8}$ inches long and $2\frac{3}{4}$ inches wide. What is the greatest area he can cover with these tiles?

3. A hotel is recarpeting a section of the lobby. Carpet covers the part of the floor as shown below in gray. How many square feet of carpeting will be needed?

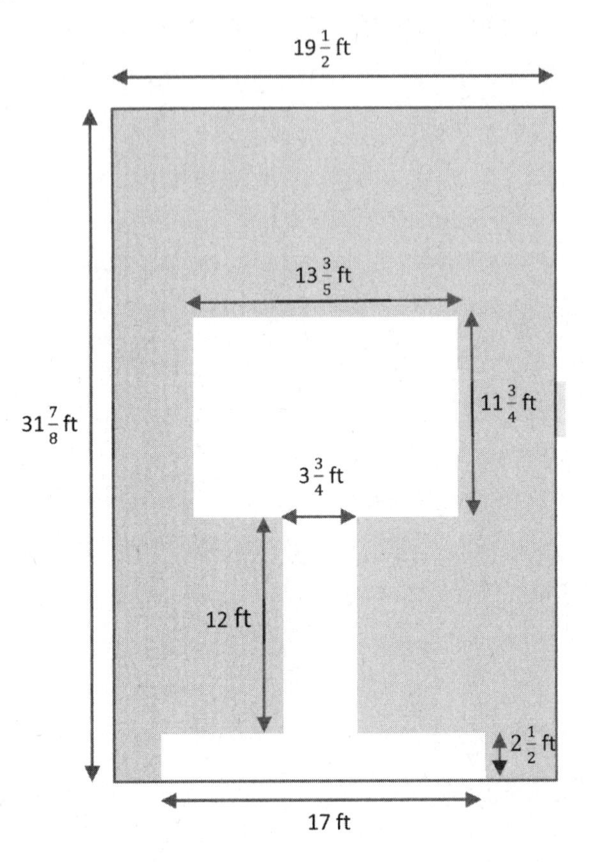

1. Sam decided to paint a wall with two windows. The gray areas below show where the windows are. The windows will not be painted. Both windows are $2\frac{1}{2}$ ft by $4\frac{1}{2}$ ft rectangles. Find the area the paint needs to cover.

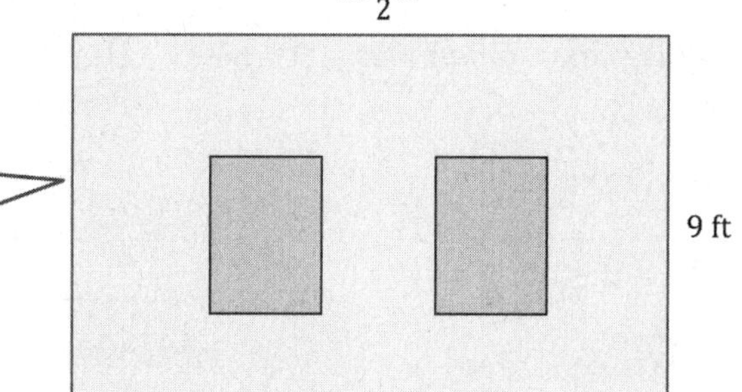

I can subtract the area of the two windows from the area of the wall to find the area that the paint needs to cover.

Area of* 1 *window:

$$2\frac{1}{2}\text{ ft} \times 4\frac{1}{2}\text{ ft}$$

$$\frac{5}{2} \times \frac{9}{2}$$

$$= \frac{45}{4}$$

$$= 11\frac{1}{4}$$

$$\text{Area} = 11\frac{1}{4}\text{ ft}^2$$

The area of 1 window is $11\frac{1}{4}$ ft^2.

Area of the wall:

$$13\frac{1}{2}\text{ ft} \times 9\text{ ft}$$

$$(13 \times 9) + \left(\frac{1}{2} \times 9\right)$$

$$= 117 + \frac{9}{2}$$

$$= 117 + 4\frac{1}{2}$$

$$= 121\frac{1}{2}$$

$$\text{Area} = 121\frac{1}{2}\text{ ft}^2$$

Area of* 2 *windows:

$$1\ \textit{unit} = 11\frac{1}{4}\text{ ft}^2$$

$$2\ \textit{units} = 2 \times 11\frac{1}{4}\text{ ft}^2$$

$$(2 \times 11) + \left(2 \times \frac{1}{4}\right)$$

$$= 22 + \frac{2}{4}$$

$$= 22\frac{1}{2}$$

$$\text{Area} = 22\frac{1}{2}\text{ ft}^2$$

I can double the area of 1 window to find the area of 2 windows. The total area is $22\frac{1}{2}$ ft^2.

I can subtract the area of the 2 windows from the area of the wall.

$$121\frac{1}{2}\text{ ft}^2 - 22\frac{1}{2}\text{ ft}^2 = 99\text{ ft}^2$$

The paint needs to cover* 99 *square feet.

2. Mason uses square tiles, some of which he cuts in half, to make the figure below. If each square tile has a side length of $3\frac{1}{2}$ inches, what is the total area of the figure?

Total tiles:

7 whole tiles + 6 half tiles = 10 tiles

I can count the tiles in the figure. There are a total of 10 tiles.

Area of 1 tile:

$$3\frac{1}{2}\text{ in} \times 3\frac{1}{2}\text{ in}$$

$$\frac{7}{2} \times \frac{7}{2}$$

$$= \frac{49}{4}$$

$$= 12\frac{1}{4}$$

$$\text{Area} = 12\frac{1}{4}\text{ in}^2$$

I can find the area of 1 square tile. $3\frac{1}{2}\text{ in} \times 3\frac{1}{2}\text{ in} = 12\frac{1}{4}\text{ in}^2$.

Area of 10 tiles:

To find the area of 10 tiles, I can multiply the area of 1 tile by 10.

$$1\ \textit{unit} = 12\frac{1}{4}\text{ in}^2$$

$$10\ \textit{units} = 10 \times 12\frac{1}{4}\text{ in}^2$$

$$(10 \times 12) + \left(10 \times \frac{1}{4}\right)$$

$$= 120 + \frac{10}{4}$$

$$= 120 + 2\frac{2}{4}$$

$$= 122\frac{1}{2}$$

The total area of the figure is $122\frac{1}{2}$ square inches.

Name ______________________________ Date ________________

1. Mr. Albano wants to paint menus on the wall of his café in chalkboard paint. The gray area below shows where the rectangular menus will be. Each menu will measure 6-ft wide and $7\frac{1}{2}$-ft tall.

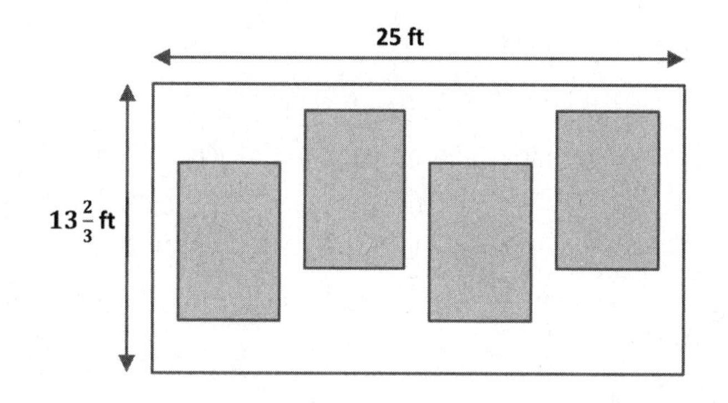

- How many square feet of menu space will Mr. Albano have?

- What is the area of wall space that is not covered by chalkboard paint?

2. Mr. Albano wants to put tiles in the shape of a dinosaur at the front entrance. He will need to cut some tiles in half to make the figure. If each square tile is $4\frac{1}{4}$ inches on each side, what is the total area of the dinosaur?

3. A-Plus Glass is making windows for a new house that is being built. The box shows the list of sizes they must make.

15 windows $4\frac{3}{4}$-ft long and $3\frac{3}{5}$-ft wide
7 windows $2\frac{4}{5}$-ft wide and $6\frac{1}{2}$-ft long

How many square feet of glass will they need?

4. Mr. Johnson needs to buy seed for his backyard lawn.

- If the lawn measures $40\frac{4}{5}$ ft by $50\frac{7}{8}$ ft, how many square feet of seed will he need to cover the entire area?

- One bag of seed will cover 500 square feet if he sets his seed spreader to its highest setting and 300 square feet if he sets the spreader to its lowest setting. How many bags of seed will he need if he uses the highest setting? The lowest setting?

1. The length of a flowerbed is 3 times as long as its width. If the width is $\frac{4}{5}$ meter, what is the area of the flowerbed?

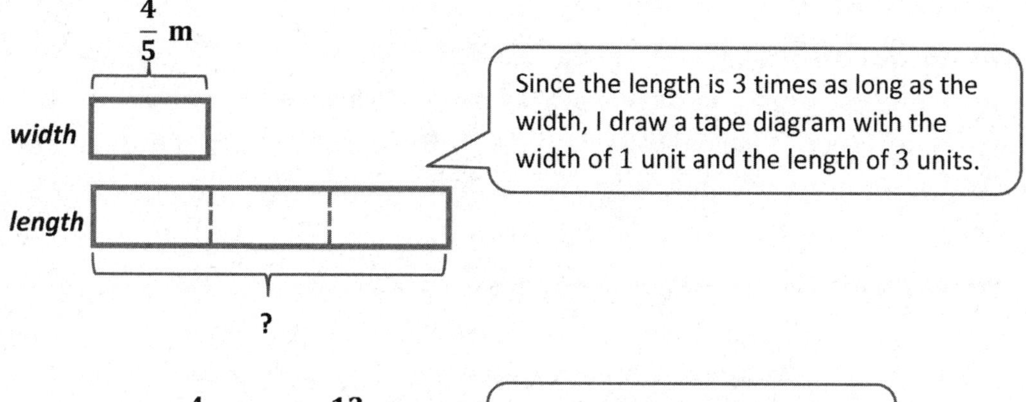

$$\frac{4}{5}\text{ m} \times 3 = \frac{12}{5}\text{ m}$$

I find the length of the flowerbed by multiplying by 3.

$$\begin{aligned}\textbf{Area} &= \textbf{length} \times \textbf{width}\\ &= \frac{12}{5}\text{ m} \times \frac{4}{5}\text{ m}\\ &= \frac{48}{25}\text{ m}^2\\ &= 1\frac{23}{25}\text{ m}^2\end{aligned}$$

I find the area of the flowerbed by multiplying the length times the width.

The flowerbed's area is $1\frac{23}{25}$ square meters.

2. Mrs. Tran grows herbs in square plots. Her rosemary plot measures $\frac{5}{6}$ yd on each side.

   a. Find the total area of the rosemary plot.

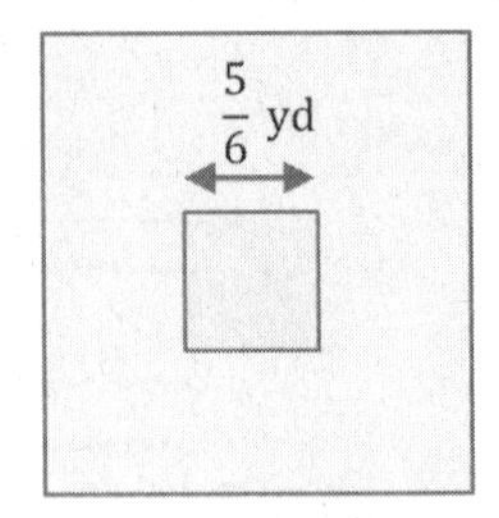

$$\textbf{Area} = \textbf{length} \times \textbf{width}$$
$$= \frac{5}{6}\ \text{yd} \times \frac{5}{6}\ \text{yd}$$
$$= \frac{25}{36}\ \text{yd}^2$$

I multiply length times width to find the area of the rosemary plot.

The total area of the rosemary plot is $\frac{25}{36}$ square yards.

   b. Mrs. Tran puts a fence around the rosemary. If the fence is 2 ft from the edge of the garden on each side, what is the perimeter of the fence?

I notice the unit here is feet, but the area I found from part (a) above was in yards.

$$\frac{5}{6}\ \text{yd} = \frac{5}{6} \times 1\ \text{yd}$$
$$= \frac{5}{6} \times 3\ \text{ft}$$
$$= \frac{15}{6}\ \text{ft}$$
$$= 2\frac{3}{6}\ \text{ft}$$
$$= 2\frac{1}{2}\ \text{ft}$$

I convert the $\frac{5}{6}$ yard into feet. The length of the rosemary plot is $2\frac{1}{2}$ feet.

One side of the fence:

$$2\frac{1}{2}\ \text{ft} + 4\ \text{ft} = 6\frac{1}{2}\ \text{ft}$$

I now find the length of one side of the fence. Since the fence is 2 feet from the edge of the garden on each side, I add 4 feet to the side of the rosemary plot, $2\frac{1}{2}$ feet. Each side of the fence is $6\frac{1}{2}$ feet long.

Perimeter of the fence:

$$6\frac{1}{2}\ \text{ft} \times 4$$
$$= (6\ \text{ft} \times 4) + \left(\frac{1}{2}\ \text{ft} \times 4\right)$$
$$= 24\ \text{ft} + \frac{4}{2}\ \text{ft}$$
$$= 24\ \text{ft} + 2\ \text{ft}$$
$$= 26\ \text{ft}$$

I multiply one side of the fence, $6\frac{1}{2}$ feet, by 4 to find the perimeter.

The perimeter of the fence is 26 feet.

Name ______________________________ Date ____________

1. The width of a picnic table is 3 times its length. If the length is $\frac{5}{6}$-yd long, what is the area of the picnic table in square feet?

2. A painting company will paint this wall of a building. The owner gives them the following dimensions:

Window A is $6\frac{1}{4}$ ft × $5\frac{3}{4}$ ft.

Window B is $3\frac{1}{8}$ ft × 4 ft .

Window C is $9\frac{1}{2}$ ft².

Door D is 4 ft × 8 ft.

What is the area of the painted part of the wall?

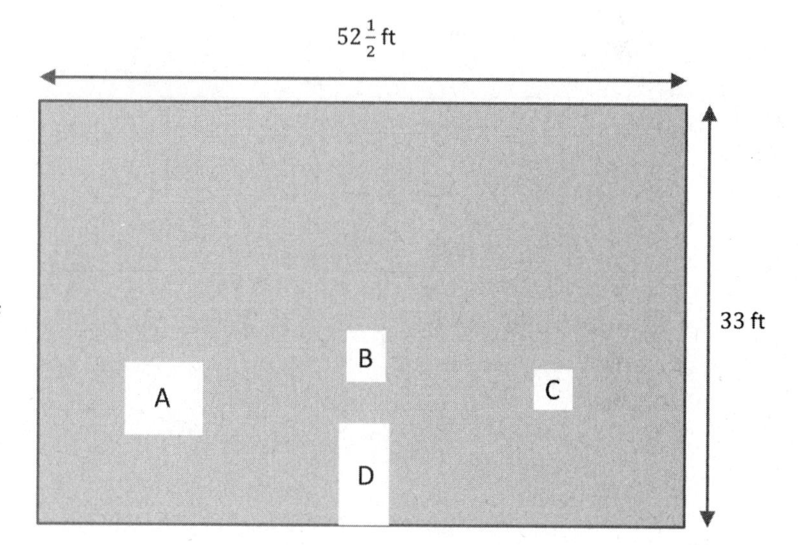

3. A decorative wooden piece is made up of four rectangles as shown to the right. The smallest rectangle measures $4\frac{1}{2}$ inches by $7\frac{3}{4}$ inches. If $2\frac{1}{4}$ inches are added to each dimension as the rectangles get larger, what is the total area of the entire piece?

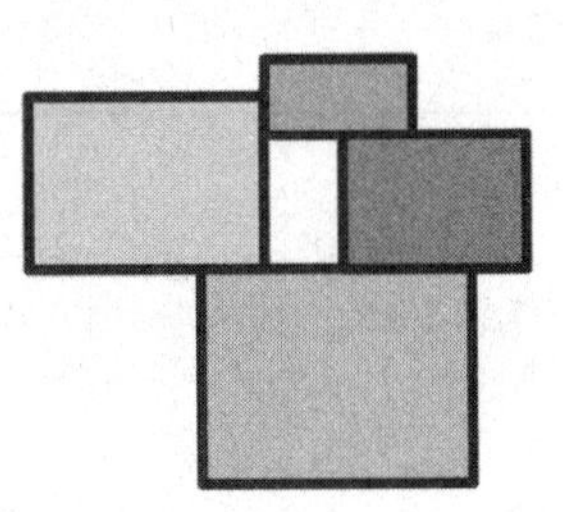

1. What are polygons with four sides called?

   Quadrilaterals

   I know that the prefix "quad" means "four."

2. What are the attributes of trapezoids?

   - ***They are quadrilaterals.***

   I know that some trapezoids with more specific attributes are commonly known as parallelograms, rectangles, squares, rhombuses, and kites. But *ALL* trapezoids are quadrilaterals with at least one set of opposite sides parallel.

   - ***They have at least one set of opposite sides parallel.***

   I know that some trapezoids have only right angles (90°), some have two acute angles (less than 90°) and two obtuse angles (more than 90° but less than 180°), and some have a combination of right, acute, and obtuse angles.

3. Use a straightedge and the grid paper to draw

   a. A trapezoid with 2 sides of equal length.

   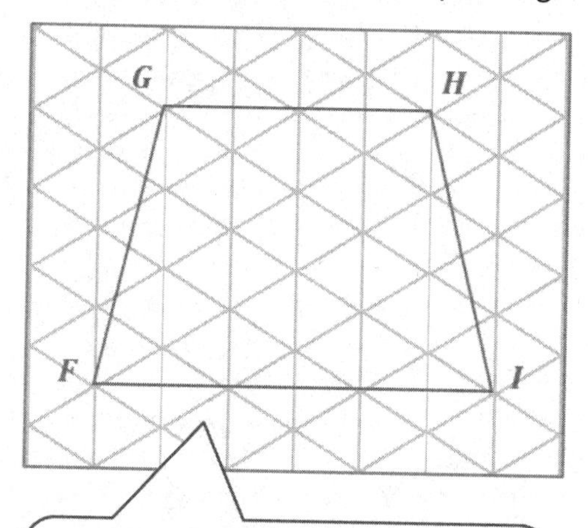

   Since this trapezoid has 2 sides of equal length ($\overline{FG}$ and $\overline{HI}$), it is called an isosceles trapezoid.

   b. A trapezoid with no sides of equal length.

   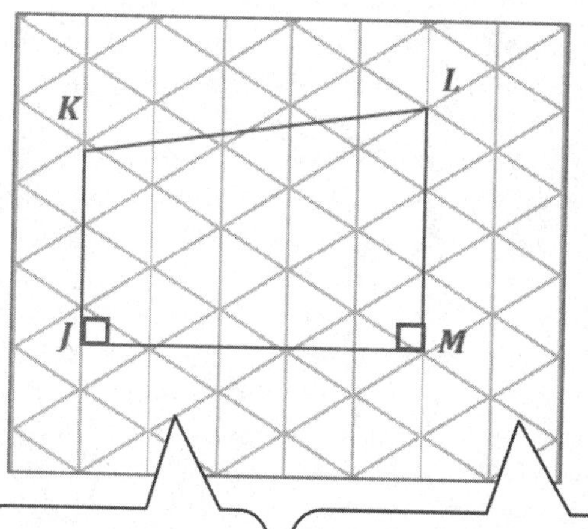

   $\angle J$ and $\angle M$ are right angles and measure 90°.

   In this trapezoid, none of the sides are equal in length.

Name ___________________________________ Date ___________________

1. Use a straightedge and the grid paper to draw:

   a. A trapezoid with exactly 2 right angles.

   b. A trapezoid with no right angles.

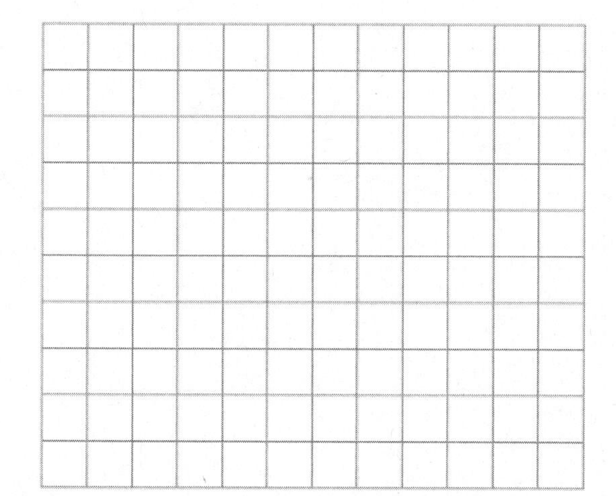

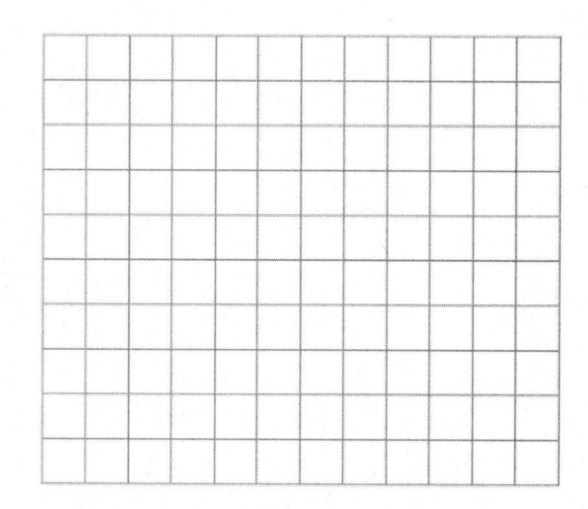

2. Kaplan incorrectly sorted some quadrilaterals into trapezoids and non-trapezoids as pictured below.

   a. Circle the shapes that are in the wrong group, and tell why they are sorted incorrectly.

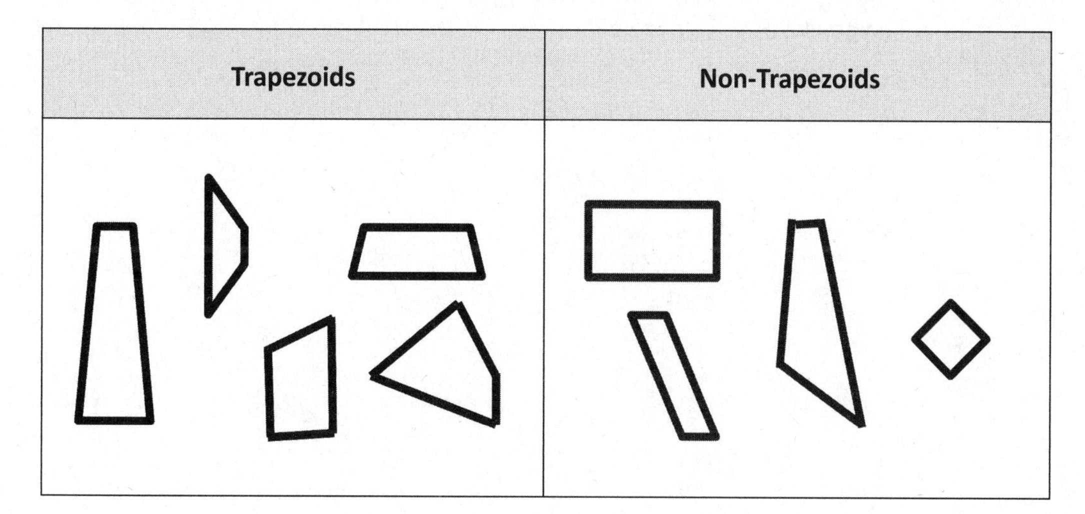

   b. Explain what tools would be necessary to use to verify the placement of all the trapezoids.

3. a. Use a straightedge to draw an isosceles trapezoid on the grid paper.

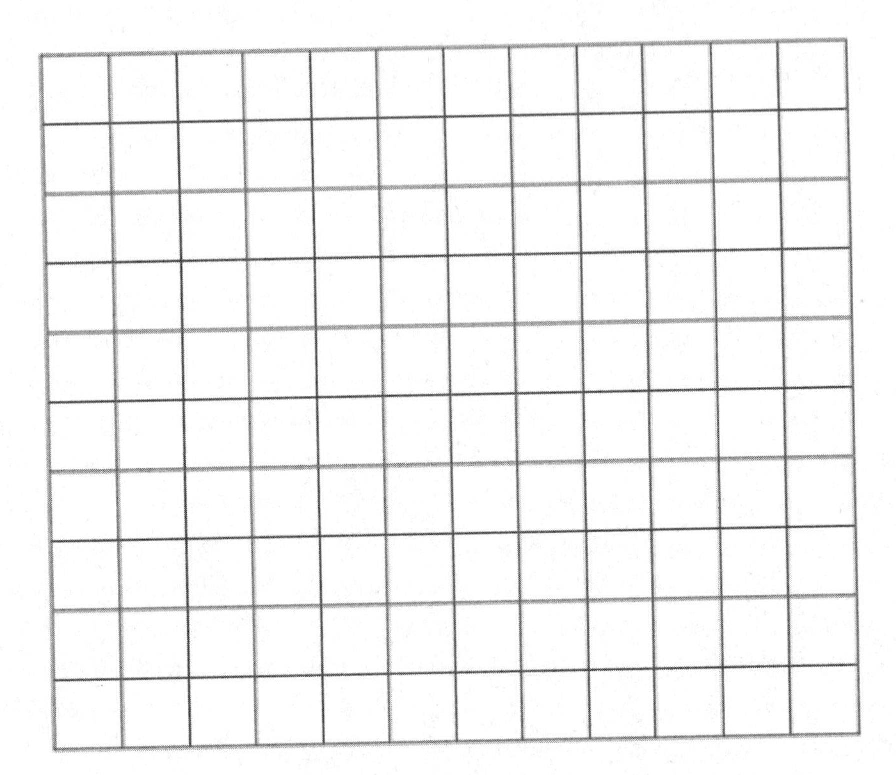

b. Why is this shape called an isosceles trapezoid?

1. Circle all of the words that could be used to name the figure below.

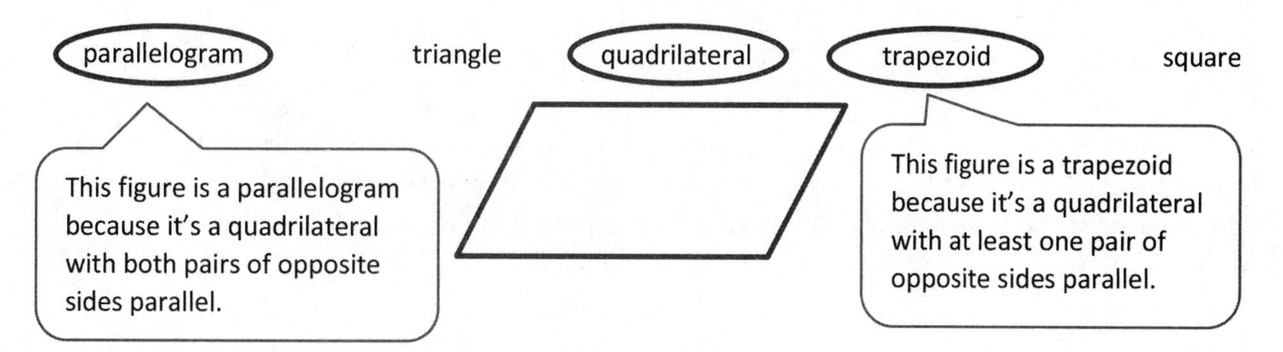

2. $HIJK$ is a parallelogram not drawn to scale.

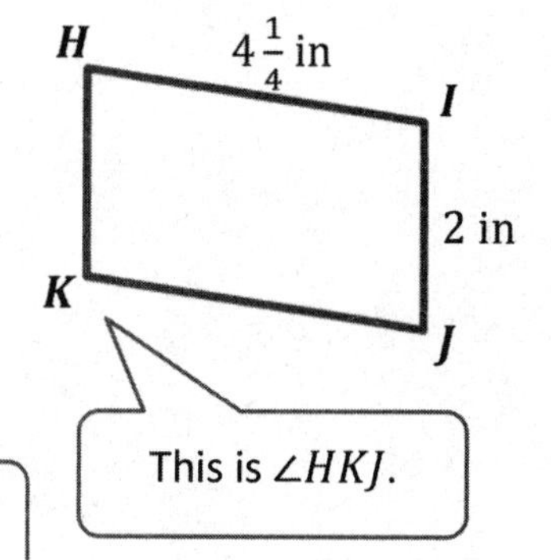

   a. Using what you know about parallelograms, give the lengths of $\overline{KJ}$ and $\overline{HK}$.

   $KJ =$ **$4\frac{1}{4}$ in**　　　　$HK =$ **2 in**

   I know that opposite sides of a parallelogram are equal in length. $HI = KJ$.

   b. $\angle HKJ = 99°$. Use what you know about angles in a parallelogram to find the measure of the other angles.

   I know that opposite angles of a parallelograms are equal in measure.

   $\angle IHK =$ **81** °　　$\angle JIH =$ **99** °　　$\angle KJI =$ **81** °

   I know that angles that are next to one another, or adjacent, add up to 180°.
   $180° - 99° = 81°$

3. $PQRS$ is a parallelogram not drawn to scale. $PR = 10$ mm and $MS =$ 4.5 mm. Give the lengths of the following segments:

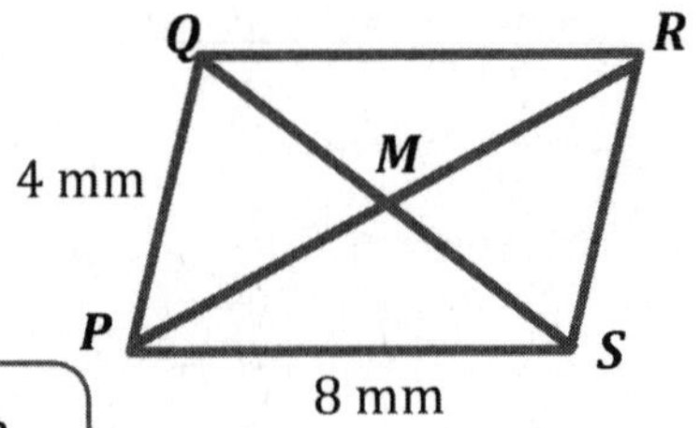

   $PM =$ **5 mm** $QS =$ **9 mm**

Q R M P S 4 mm 8 mm

I know that the diagonals of a parallelogram bisect, or cut one another in two equal parts. So the length of $\overline{PM}$ is equal to half the length of $\overline{PR}$.

Name ____________________________________ Date ______________

1. $\angle A$ measures 60°.

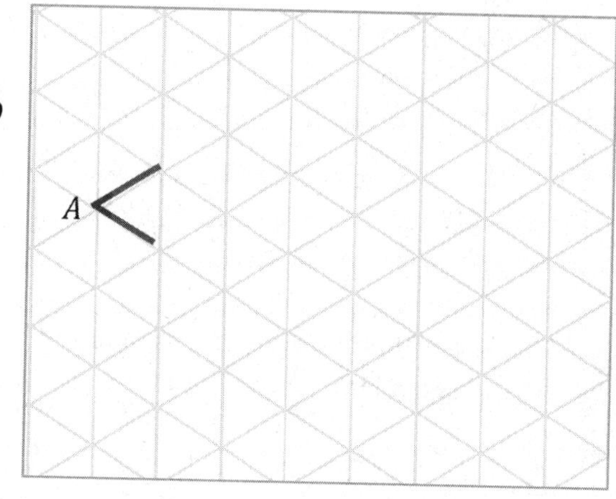

   a. Extend the rays of $\angle A$, and draw parallelogram $ABCD$ on the grid paper.

   b. What are the measures of $\angle B$, $\angle C$, and $\angle D$?

2. $WXYZ$ is a parallelogram not drawn to scale.

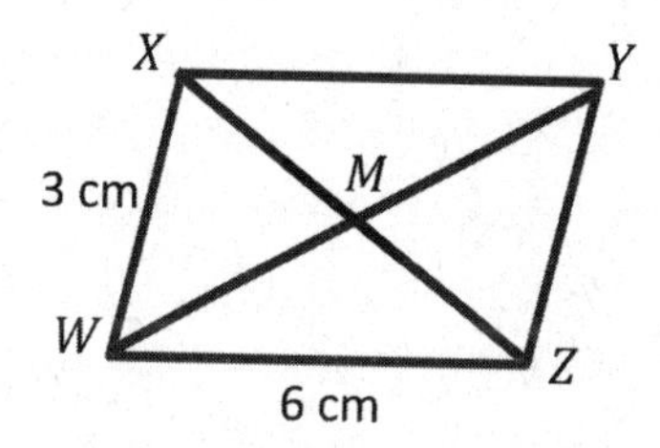

   a. Using what you know about parallelograms, give the measure of sides XY and YZ.

   b. $\angle WXY = 113°$. Use what you know about angles in a parallelogram to find the measure of the other angles.

$\angle XYZ$ = ________° $\angle YZW$ = ________° $\angle ZWX$ = ________°

3. Jack measured some segments in Problem 2. He found that $\overline{WY} = 8$ cm and $\overline{MZ} = 3$ cm.

   Give the lengths of the following segments:

   WM = ________ cm MY = ________ cm

   XM = ________ cm XZ = ________ cm

4. Using the properties of shapes, explain why all parallelograms are trapezoids.

5. Teresa says that because the diagonals of a parallelogram bisect each other, if one diagonal is 4.2 cm, the other diagonal must be half that length. Use words and pictures to explain Teresa's error.

1. What is the definition of a rhombus? Draw an example.

A rhombus is a quadrilateral (a shape with 4 sides) with all sides equal in length.

One example of a rhombus looks like this:

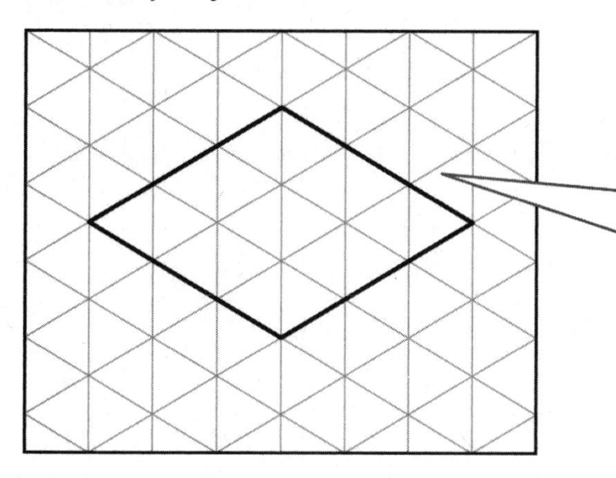

My rhombus looks like a diamond, but I could have drawn it other ways, too. As long as it is a quadrilateral with 4 sides of equal length, it is a rhombus.

2. What is the definition of a rectangle? Draw an example.

A rectangle is a quadrilateral with four right (90 degree) angles.

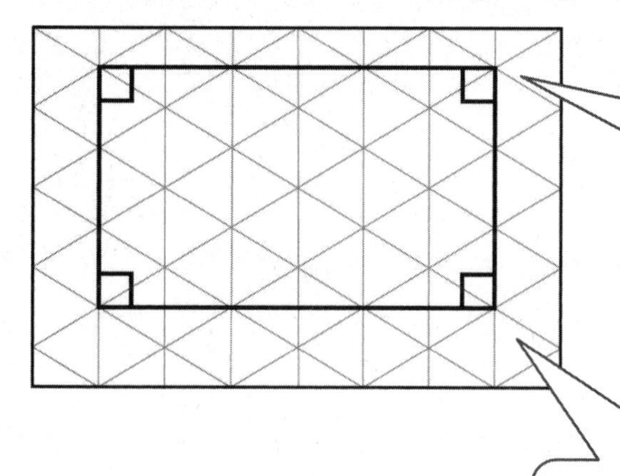

My rectangle has 2 long sides and 2 short sides, but I could have drawn it other ways, too. As long as it is a quadrilateral with right angles, it is a rectangle.

The boxes in the corners of my rectangle show that all the angles are 90 degrees.

Name ______________________________ Date ______________

1. Use the grid paper to draw.

a. A rhombus with no right angles

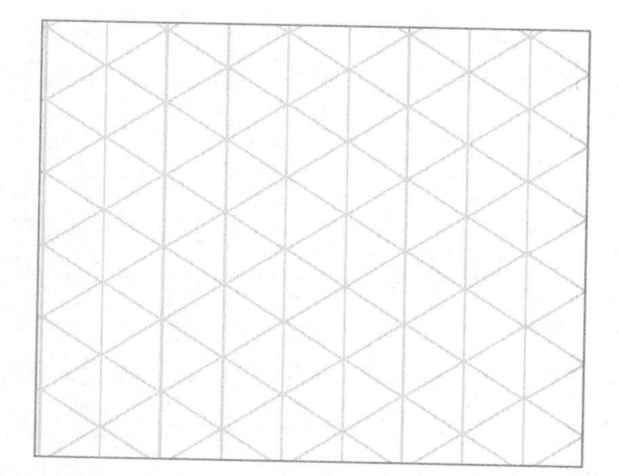

b. A rhombus with 4 right angles

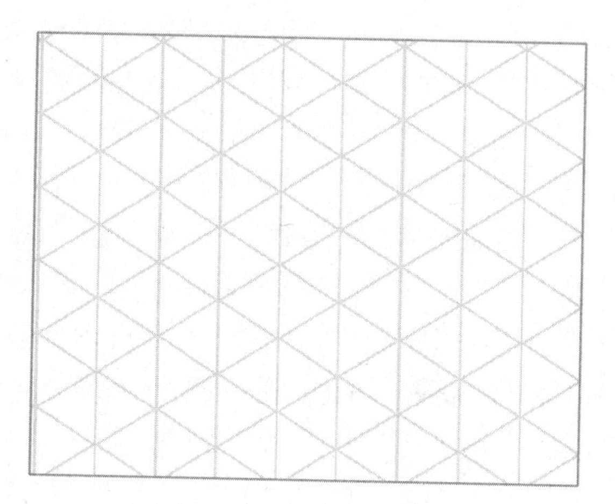

c. A rectangle with not all sides equal

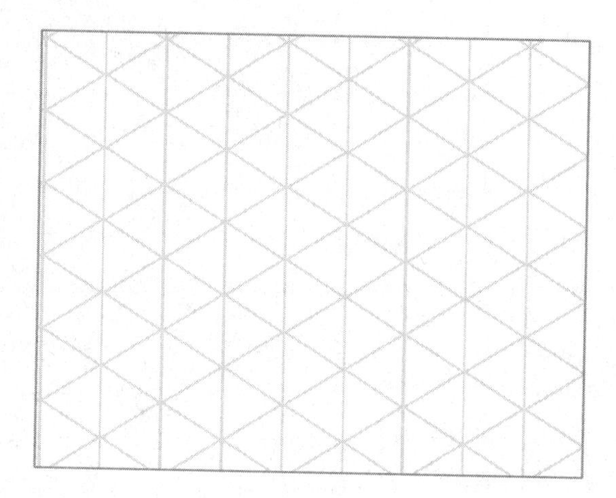

d. A rectangle with all sides equal

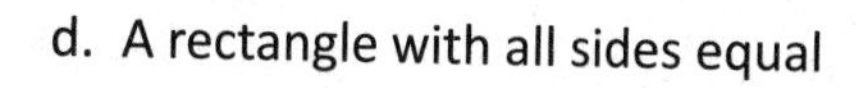

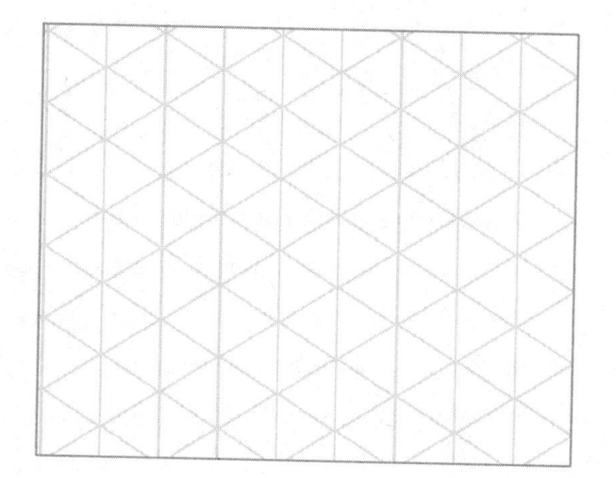

2. A rhombus has a perimeter of 217 cm. What is the length of each side of the rhombus?

3. List the properties that all rhombuses share.

4. List the properties that all rectangles share.

1. What are the attributes of a square? Draw an example.

The attributes of a square are

- ***Four sides that are equal in length (same as a rhombus)***
- ***Four right angles (same as a rectangle)***
- ***A square is a type of rhombus and a type of rectangle!***

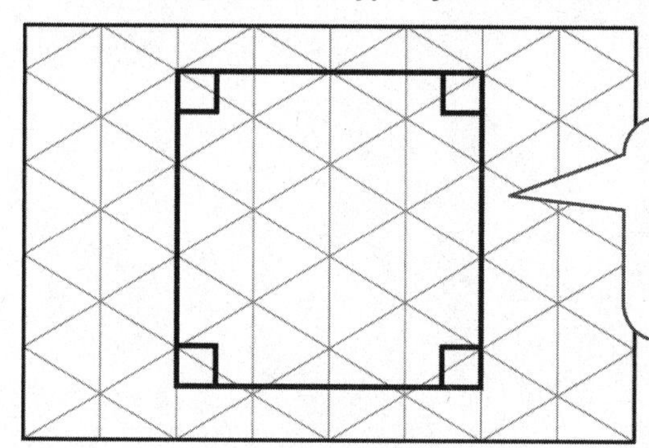

This is a square.
It is also a rhombus because it has 4 sides of equal length.
It is also a rectangle because it has 4 right angles.

2. What are the attributes of a kite? Draw an example.

The attributes of a kite are

- ***A quadrilateral in which 2 consecutive (next to each other) sides are equal in length.***
- ***The other 2 side lengths are equal to one another as well.***

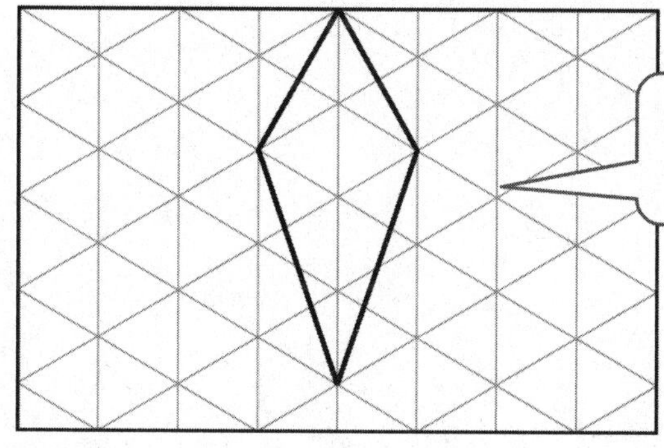

The 2 sides on "top" are equal in length, and the 2 sides on the "bottom" are equal in length.

3. Is the kite you drew in Problem 2, a parallelogram? Why or why not?

No, the kite I drew is not a parallelogram. A parallelogram must have both sets of opposite sides parallel. There are no parallel sides in my kite. The only time a kite is a parallelogram is when the kite is a square or a rhombus.

Name ______________________________ Date ______________

1. a. Draw a kite that is not a parallelogram on the grid paper.

   b. List all the properties of a kite.

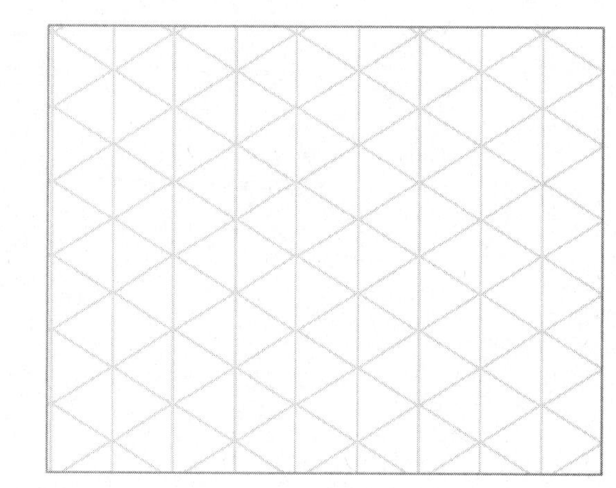

   c. When can a parallelogram also be a kite?

2. If rectangles must have right angles, explain how a rhombus could also be called a rectangle.

3. Draw a rhombus that is also a rectangle on the grid paper.

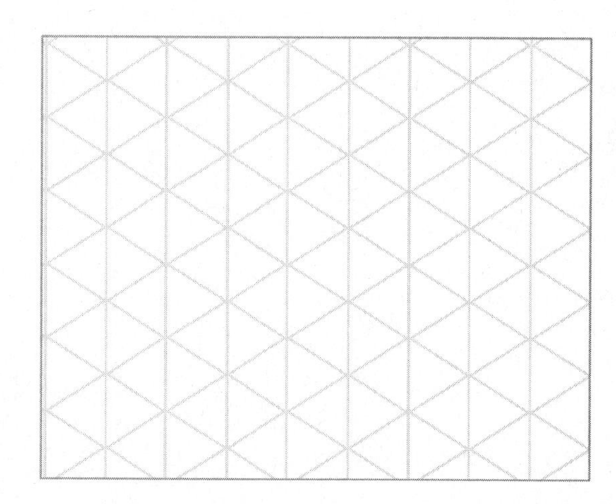

4. Kirkland says that figure $EFGH$ below is a quadrilateral because it has four points in the same plane and four segments with no three endpoints collinear. Explain his error.

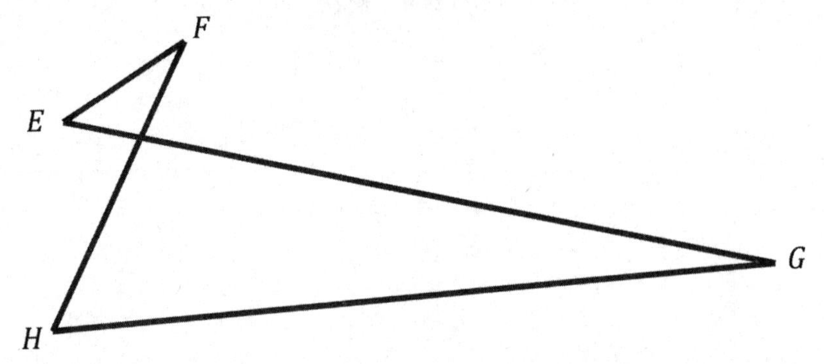

1. Fill in the table below.

Shape	Defining Attributes
Trapezoid	• Quadrilateral • Has at least one pair of parallel sides
Parallelogram	• ***A quadrilateral in which both pairs of opposite sides are parallel***
Rectangle	• A quadrilateral with 4 right angles
Rhombus	• A quadrilateral with all sides of equal length
Square	• A rhombus with four 90° angles • A rectangle with 4 equal sides
Kite	• ***Quadrilateral with 2 consecutive sides of equal length*** • ***Has 2 remaining sides of equal length***

2. $TUVW$ is a square with an area of 81 cm^2, and $UB = 6.36 \text{ cm}$. Find the measurements using what you know about the properties of squares.

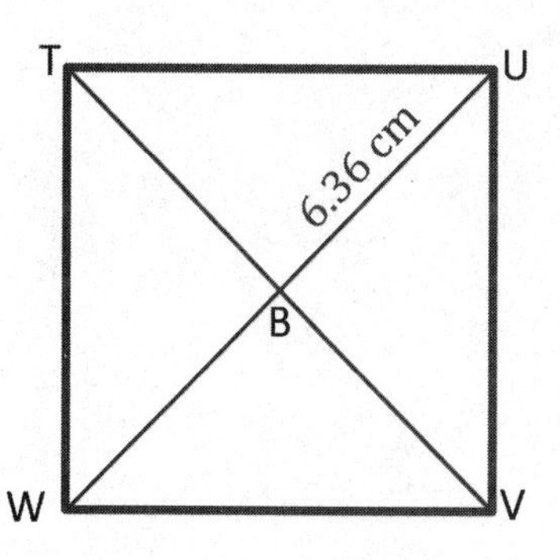

a. $UW =$ **12.72** cm

Diagonals of a square bisect each other, so $\overline{UB}$ and $\overline{BW}$ are equal in length. $6.36 + 6.36 = 12.72$

b. $TV = UW = 12.72 \text{ cm}$

I know that in a square the diagonals are equal in length.

c. Perimeter = **36** cm

I know that in a square every side length is equal, so I need to think about what times itself is equal to 81. I know that 9×9 is 81, so each side is 9 cm. Since there are 4 equal sides, I can multiply 9×4 to get the perimeter.

d. $m\angle TUV =$ **90** °

I know every angle in a square must be 90° because it is a defining attribute of a square.

Name ______________________________ Date ______________

1. Follow the flow chart, and put the name of the figure in the boxes.

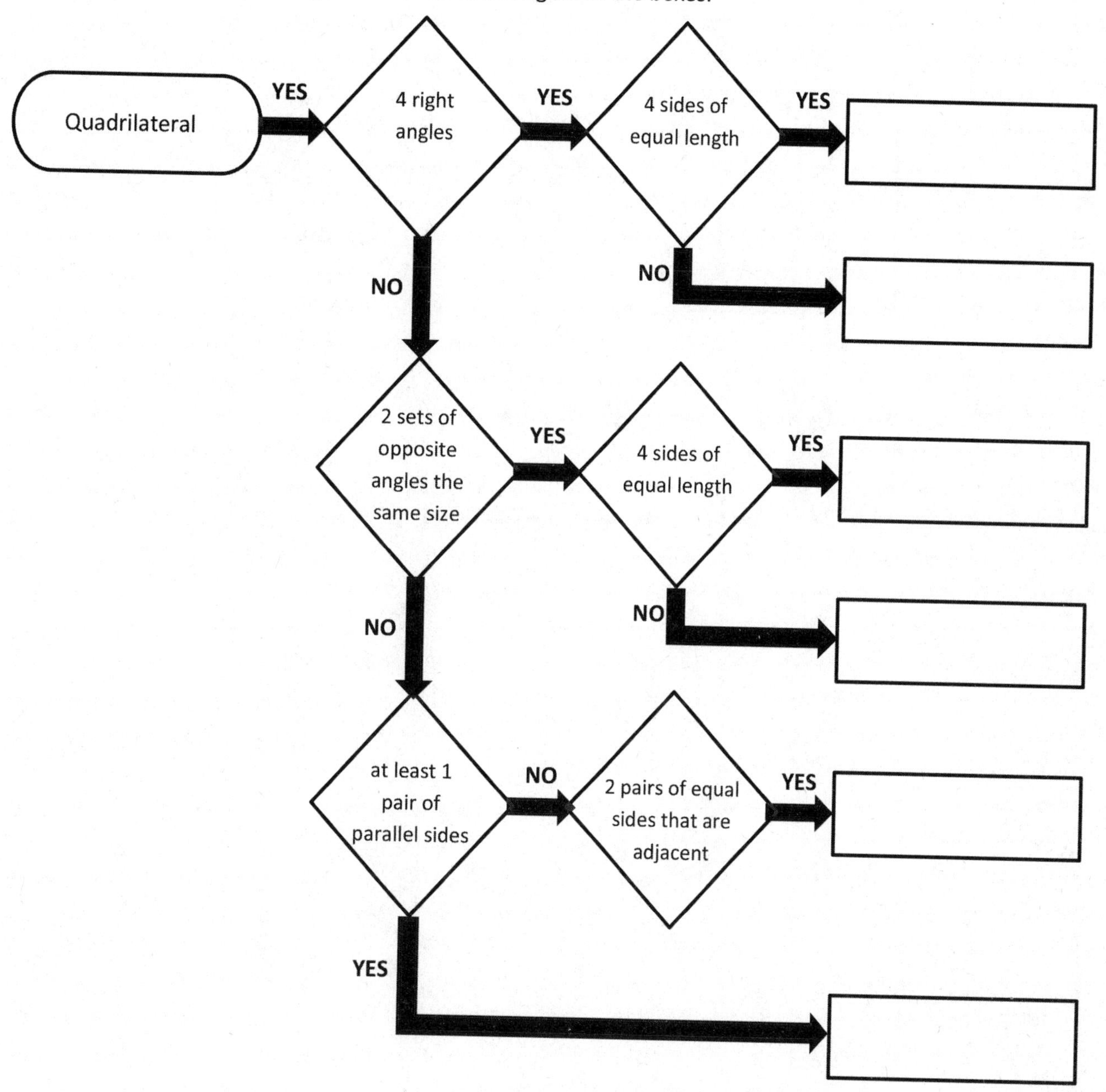

2. $SQRE$ is a square with an area of 49 cm², and $RM = 4.95$ cm. Find the measurements using what you know about the properties of squares.

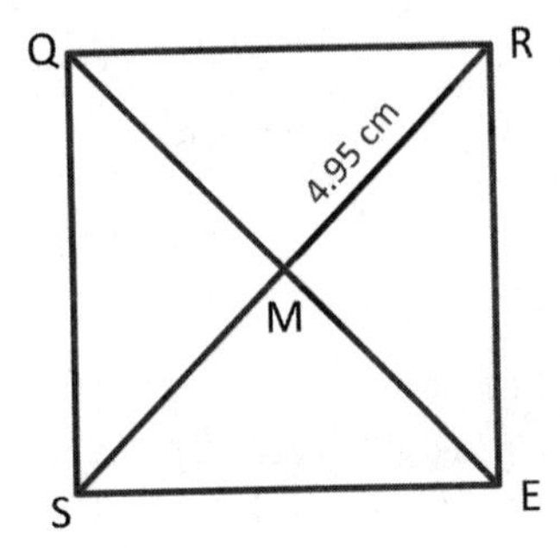

a. RS = _________ cm

b. QE = _________ cm

c. Perimeter = _________ cm

d. $m\angle QRE$ = _________ °

e. $m\angle RMQ$ = _________ °

Finish each sentence below by writing "sometimes" or "always" in the first blank, and then state the reason why. Sketch an example of each statement in the space to the right.

a. A rectangle is ***sometimes*** a square because ***a rectangle has 4 right angles, and a square is a special type of rectangle with 4 equal sides.***

This is a rectangle. It is ***not*** a square because all 4 sides are not equal in length.

b. A square is ***always*** a rectangle because ***a rectangle is a parallelogram with 4 right angles. A square is a rectangle with 4 equal sides.***

This is a square and a rectangle because it has 4 right angles and 4 equal sides.

c. A rectangle is ***sometimes*** a kite because ***a square fits the definition of a kite and rectangle. A kite has two pairs of sides that are equal, which is the same as a square.***

This is a kite, a square, and a rectangle. It has 4 right angles and 2 sets of consecutive sides equal in length.

d. A rectangle is ***always*** a parallelogram because ***it has two pairs of parallel sides.***

All rectangles can also be called parallelograms.

e. A square is ***always*** a trapezoid because ***it has at least one pair of parallel sides.***

This square, and all squares, has 2 pairs of opposite sides that are parallel. All squares can also be called trapezoids.

f. A trapezoid is ***sometimes*** a parallelogram because ***a trapezoid has to have at least one pair of parallel sides, but it could have two pairs, which fits the definition of a parallelogram.***

This figure is a trapezoid but ***not*** a parallelogram. It only has 1 pair of opposite sides parallel. (The "top" and "bottom" sides are parallel.)

Name ______________________________ Date ________________

1. Answer the questions by checking the box.

	Sometimes	Always
a. Is a square a rectangle?		
b. Is a rectangle a kite?		
c. Is a rectangle a parallelogram?		
d. Is a square a trapezoid?		
e. Is a parallelogram a trapezoid?		
f. Is a trapezoid a parallelogram?		
g. Is a kite a parallelogram?		

h. For each statement that you answered with *sometimes*, draw and label an example that justifies your answer.

2. Use what you know about quadrilaterals to answer each question below.

a. Explain when a trapezoid is not a parallelogram. Sketch an example.

b. Explain when a kite is not a parallelogram. Sketch an example.

Grade 5
Module 6

1. Answer the following questions using number line P, below.

   The origin is always zero.

   a. What is the coordinate, or the distance from the origin, of the ⬟ ?

   20

   The coordinate tells the distance from the zero to the shape on the number line.

   b. What is the coordinate of ▲ ?

   25

   c. What is the coordinate at the midpoint between (and ⬟ ?

   15

   The distance from the moon to the pentagon is 10 units, so the midpoint will be 5 units from each shape.

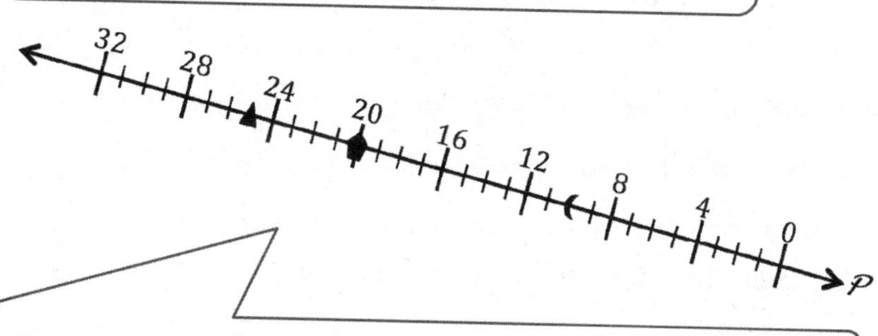

This number line increases from right to left. Number lines can go in any direction.

2. Use the number line to answer the questions.

   The first tick mark is 0, and the second is 0.4. The distance between tick marks is 0.4, or $\frac{4}{10}$.

   a. Plot P so its distance is $\frac{2}{10}$ from the origin.

   b. Plot Q 12 tenths farther from the origin than point P.

   12 tenths more than 2 tenths is 14 tenths, or 1.4.

   c. Plot R so its distance is 1 closer to the origin than point Q.

   d. What is the distance from P to R?

   The distance from P to R is $\mathbf{0.2}$.

   I can think of 1 as 10 tenths.

0
P
R 0.4
0.8
1.2
Q
1.6

3. Number line L shows 18 units. Use number line L, below, to answer the questions.

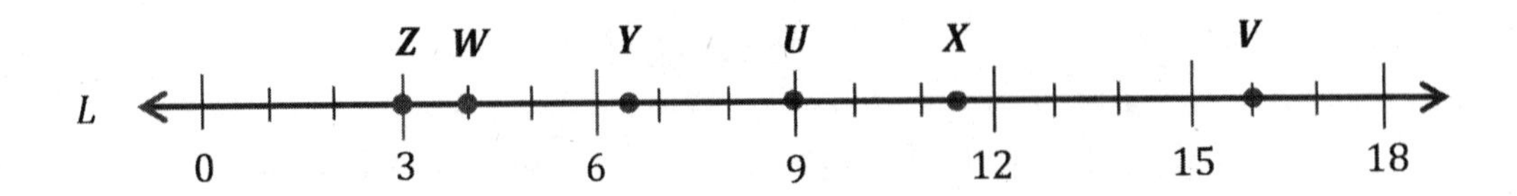

a. Plot a point at 3. Label it Z.

The units are one, and they are indicated by the tick marks on the number line.

b. Label point Y at $6\frac{1}{2}$.

"Closer to the origin" means I have to move to the left along this number line.

c. Plot a point X that is 5 units farther from zero than point Y.

d. Plot point W that is $\frac{5}{2}$ units closer to the origin than point Y. What is the coordinate of point W?

The coordinate of point W is 4.

e. What is the coordinate of the point that is 4.5 units farther from the origin than point X? Label this point V.

The coordinate of point V is 16.

$11\frac{1}{2} + 4\frac{1}{2} = 16$

f. Label point U midway between point Y and point X. What is the coordinate of this point?

The coordinate midway between points Y and X is 9.

4. A pirate buried stolen treasure in a vacant lot. He made a note that he buried the treasure 15 feet from the only tree on the lot. Later he could not find the treasure. What did he do wrong?

He did not indicate what direction from the tree he buried the treasure. If he just says fifteen feet from the tree, he'd have to dig a circle around the tree to find the treasure.

Name ______________________________ Date ____________

1. Answer the following questions using number line q below.

   a. What is the coordinate, or the distance from the origin, of the 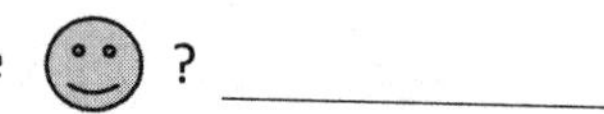 ? ____________

   b. 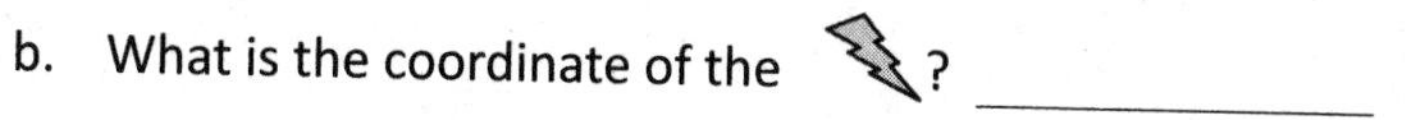 What is the coordinate of the ? ____________

   c. What is the coordinate of the 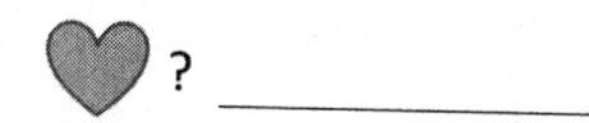 ? ____________

   d. What is the coordinate at the midpoint of the and the ? ____________

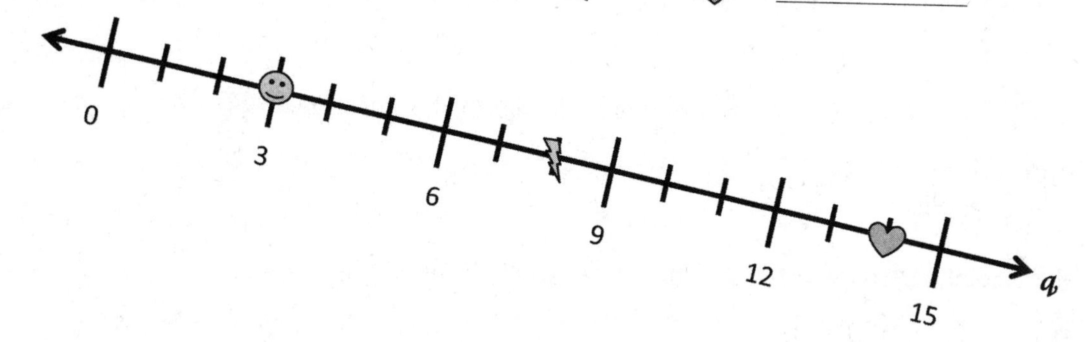

2. Use the number lines to answer the questions.

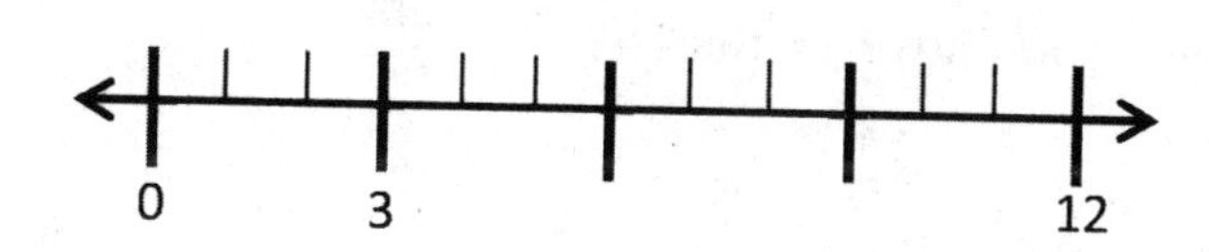

Plot T so that its distance from the origin is 10.

0 P 3

Plot M so that its distance is $\frac{11}{4}$ from the origin. What is the distance from P to M?

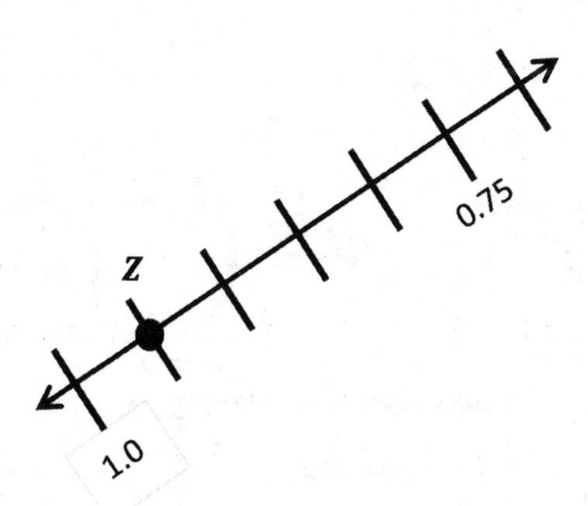

Plot a point that is 0.15 closer to the origin than Z.

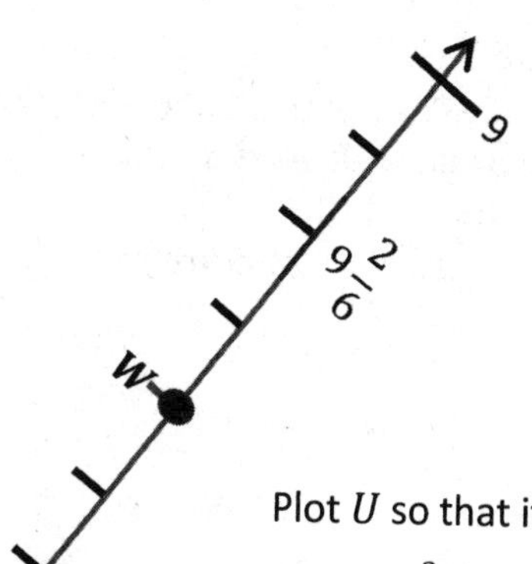

Plot U so that its distance from the origin is $\frac{3}{6}$ less than that of W.

3. Number line k shows 12 units. Use number line k below to answer the questions.

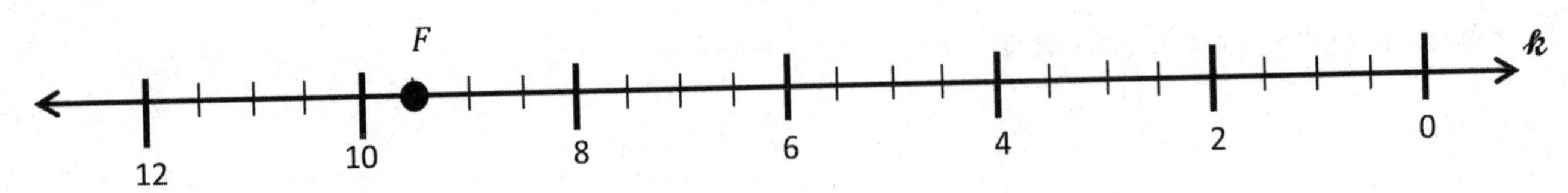

a. Plot a point at 1. Label it A.

b. Label a point that lies at $3\frac{1}{2}$ as B.

c. Label a point, C, whose distance from zero is 8 units farther than that of B.

The coordinate of C is __________.

d. Plot a point, D, whose distance from zero is $\frac{6}{2}$ less than that of B.

The coordinate of D is __________.

e. What is the coordinate of the point that lies $\frac{17}{2}$ farther from the origin than D?

Label this point E.

f. What is the coordinate of the point that lies halfway between F and D?

Label this point G.

4. Mr. Baker's fifth-grade class buried a time capsule in the field behind the school. They drew a map and marked the location of the capsule with an ✖ so that his class can dig it up in ten years. What could Mr. Baker's class have done to make the capsule easier to find?

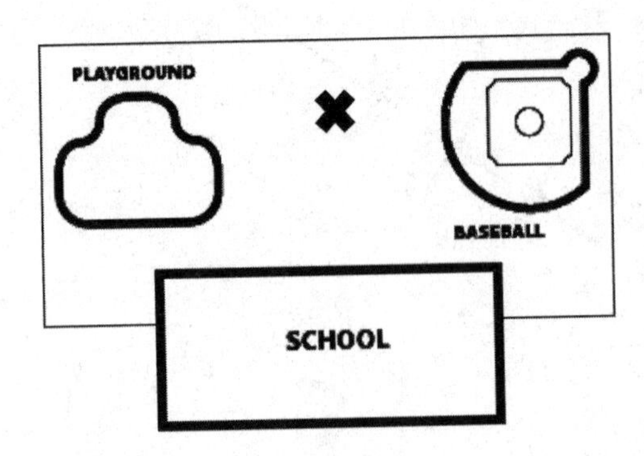

1. Use a set square to draw a line perpendicular to the x-axis through point R. Label the new line as the y-axis.

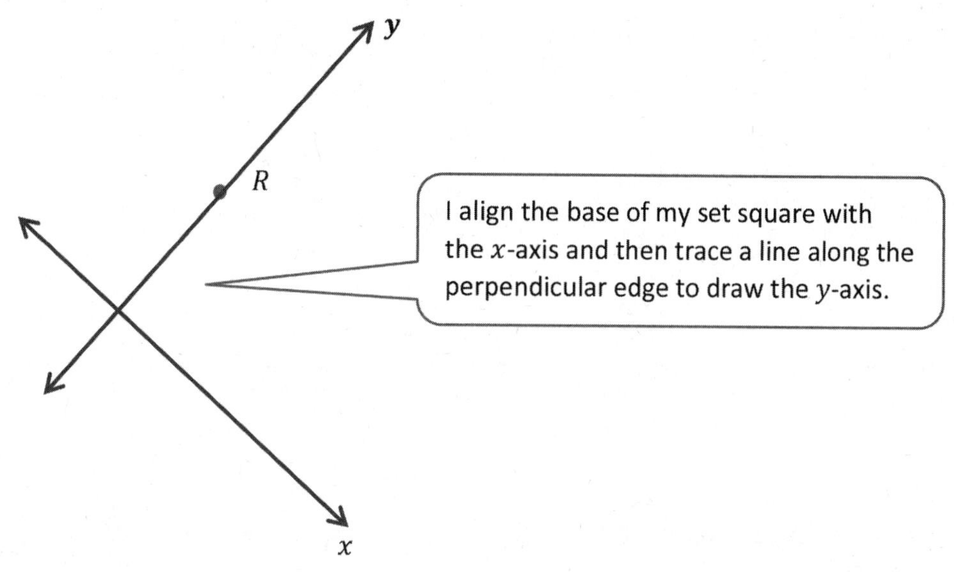

2. Use the perpendicular lines below to create a coordinate plane. Mark 6 units on each axis, and label them as fractions.

I chose fractional units of $\frac{1}{2}$, but I could have chosen any fractional unit.

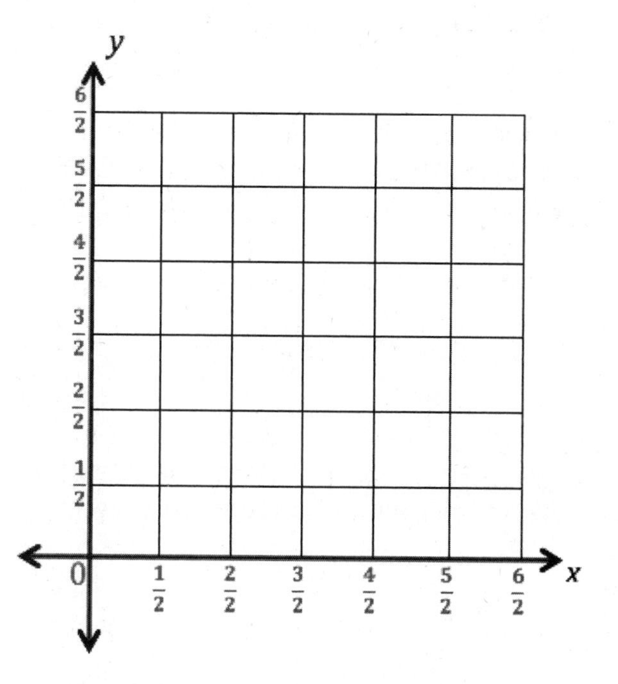

3. Use the coordinate plane to answer the following.

x-coordinate	y-coordinate	Shape
$1\frac{1}{2}$	0	***circle***
4.5	1.5	***trapezoid***
2	3	***flag***
3	4	***square***

$1\frac{1}{2}$ is not one of the numbers on the x-axis, but I know that $1\frac{1}{2}$ falls halfway between 1 and 2.

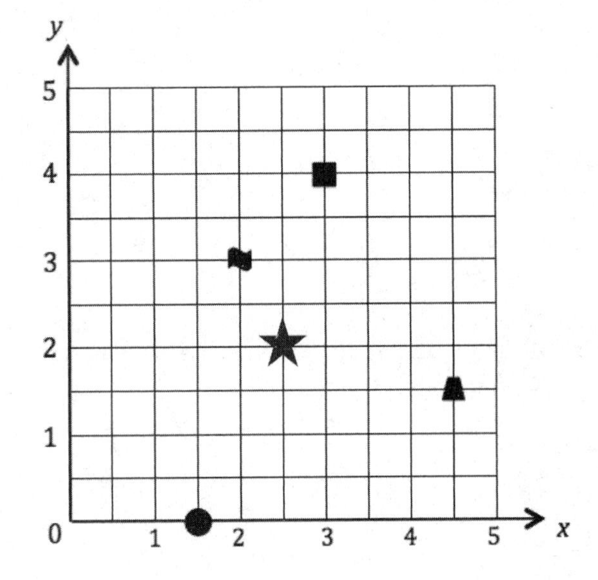

a. Name the shape at each location.

b. What shape is 3 units from the x-axis?

The flag is 3 units from the x-axis.

c. Which shape has a y-coordinate of 3?

The flag has a y-coordinate of 3.

Problems 3(b) and 3(c) are asking the same question in different ways.

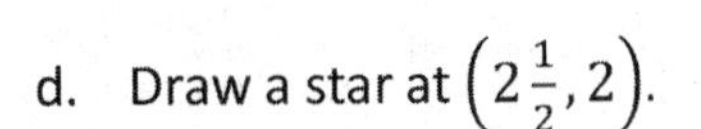

d. Draw a star at $\left(2\frac{1}{2}, 2\right)$.

The numbers in the parentheses are *coordinate pairs.* Coordinate pairs are written in parentheses with a comma separating the two coordinates. The *x-coordinate* is given first.

Name ______________________________ Date ______________

1.

a. Use a set square to draw a line perpendicular to the x-axis through point P. Label the new line as the y-axis.

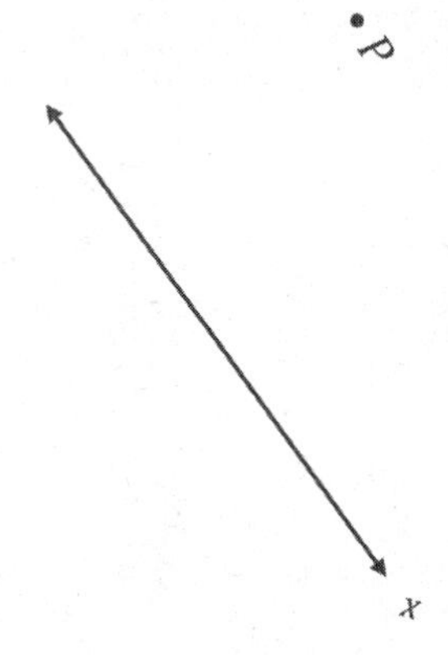

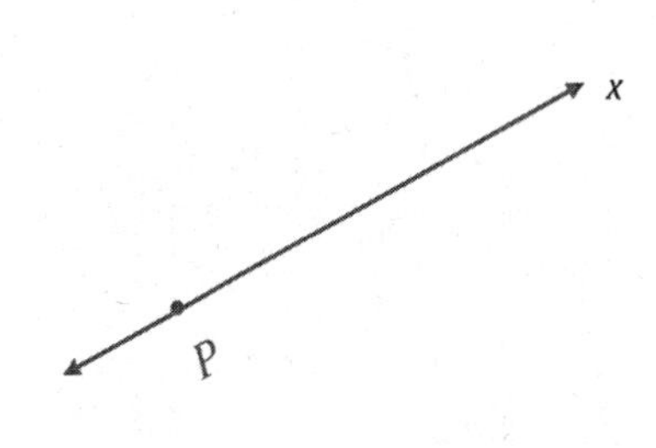

b. Choose one of the sets of perpendicular lines above, and create a coordinate plane. Mark 5 units on each axis, and label them as whole numbers.

2. Use the coordinate plane to answer the following.

a. Name the shape at each location.

x-coordinate	y-coordinate	Shape
2	4	
5	4	
1	5	
5	1	

b. Which shape is 2 units from the x-axis?

c. Which shape has the same x- and y-coordinate?

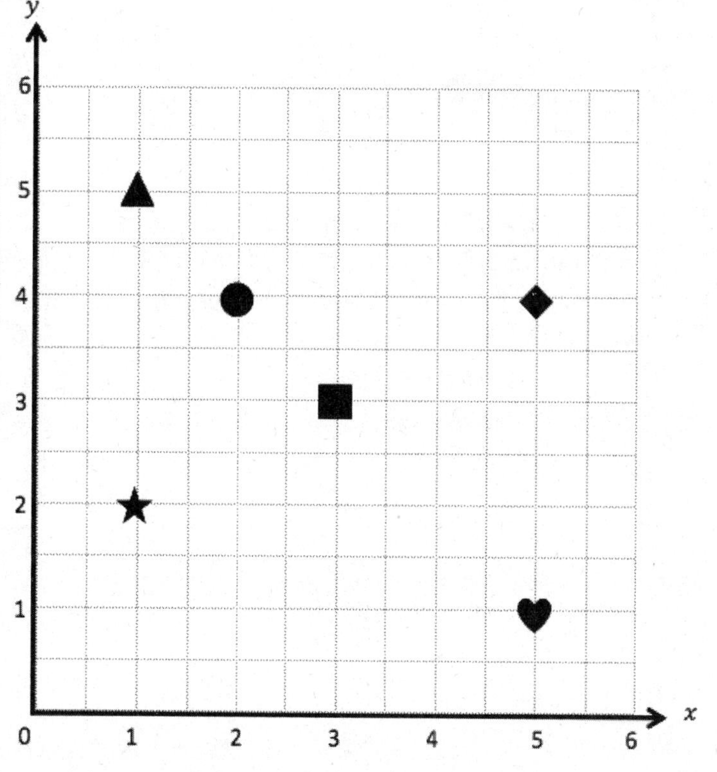

3. Use the coordinate plane to answer the following.

   a. Name the coordinates of each shape.

Shape	x-coordinate	y-coordinate
Moon		
Sun		
Heart		
Cloud		
Smiley Face		

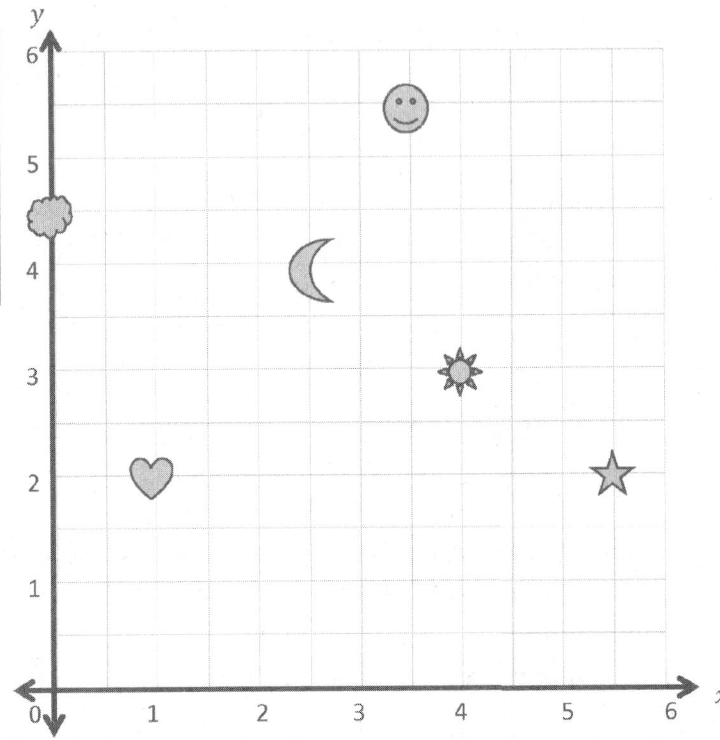

   b. Which 2 shapes have the same y-coordinate?

   c. Plot an X at (2, 3).

   d. Plot a square at $(3, 2\frac{1}{2})$.

   e. Plot a triangle at $(6, 3\frac{1}{2})$.

4. Mr. Palmer plans to bury a time capsule 10 yards behind the school. What else should he do to make naming the location of the time capsule more accurate?

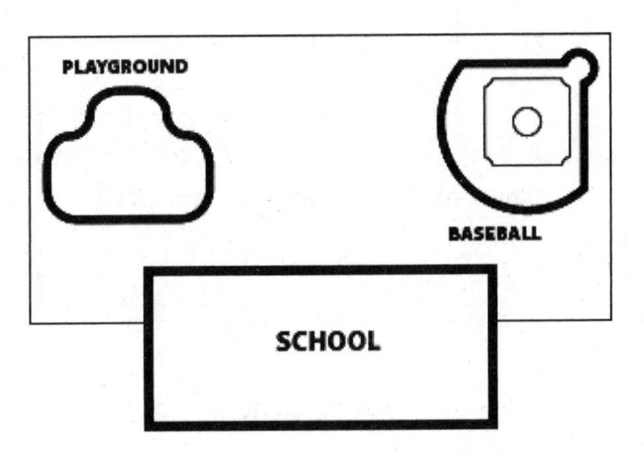

The y-axis is a vertical line. The x-axis is a horizontal line.

The origin, or $(0, 0)$, is where the x- and y-axes meet.

1. Use the grid below to complete the following tasks.
   a. Construct a y-axis that passes through points A and B. Label this axis.
   b. Construct an x-axis that is perpendicular to the y-axis that passes through points A and M.
   c. Label the origin.
   d. The x-coordinate of point W is $2\frac{3}{4}$. Label the whole numbers along the x-axis.
   e. Label the whole numbers along the y-axis.

The y-axis must be labeled the same way as the x-axis. On the x-axis, the distance between grid lines is $\frac{1}{4}$. I can use the same units for the y-axis.

I find point W on the coordinate plane. I can trace down with my finger to locate this spot on the x-axis. I count back to 0 and see that each line on the grid is $\frac{1}{4}$ more than the previous line.

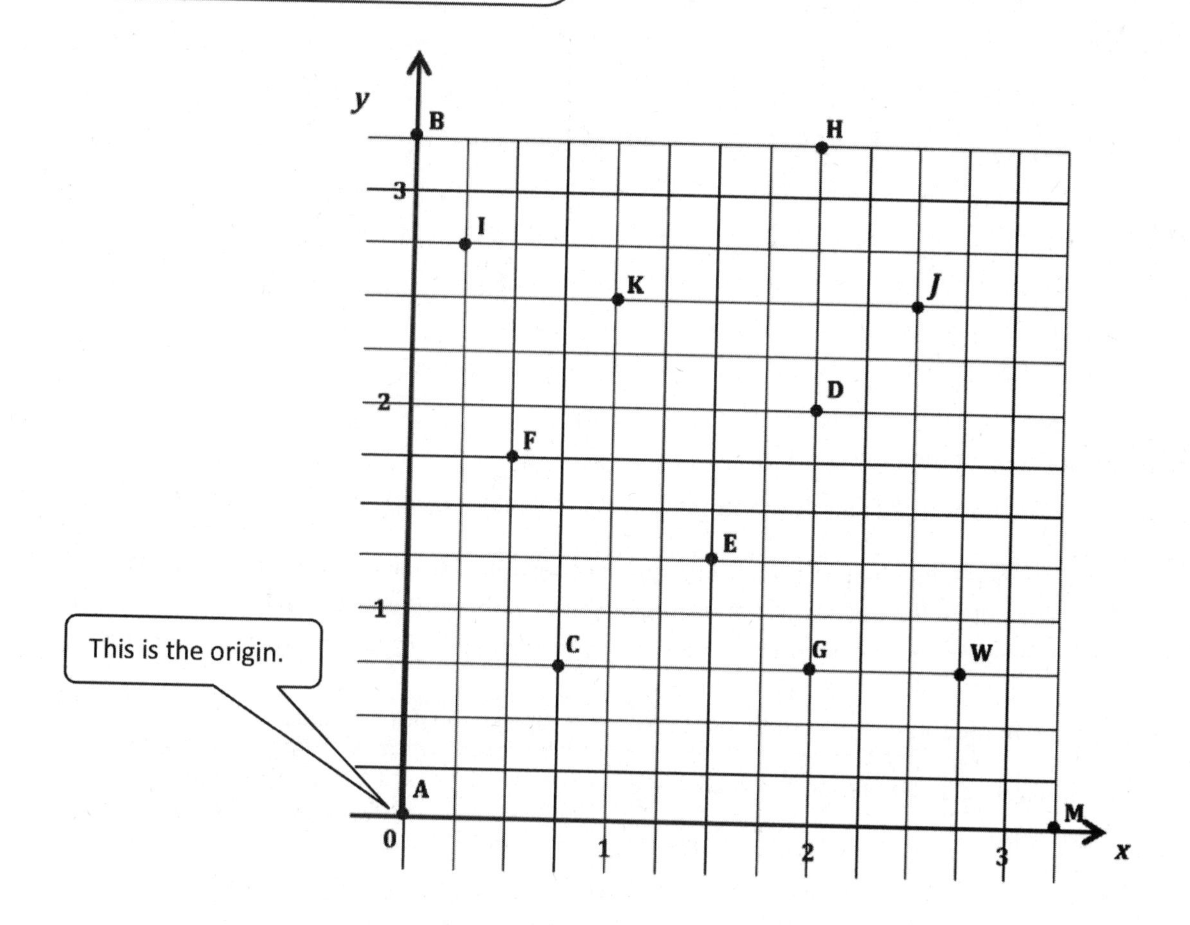

2. For the following problems, consider all the points on the previous page.

   a. Identify all the points that have a y-coordinate of $\frac{3}{4}$.

   C, G, and W

   I look for all of the points that are $\frac{3}{4}$ units from the x-axis.

   b. Identify all the points that have an x-coordinate of 2.

   G, D, and H

   I look for points that are 2 units from the y-axis.

   c. Name the point, and write the coordinate pair that is $2\frac{1}{2}$ units above the x-axis and 1 unit to the right of the y-axis.

   $\boldsymbol{K\left(1, 2\frac{1}{2}\right)}$

   d. Which point is located $1\frac{1}{4}$ units from the x-axis? Give its coordinates.

   $\boldsymbol{E\left(1\frac{1}{2}, 1\frac{1}{4}\right)}$

   e. Which point is located $\frac{1}{4}$ units from the y-axis? Give its coordinates.

   $\boldsymbol{I\left(\frac{1}{4}, 2\frac{3}{4}\right)}$

   f. Give the coordinates for point C.

   $\boldsymbol{\left(\frac{3}{4}, \frac{3}{4}\right)}$

   g. Plot a point where both coordinates are the same. Label the point J, and give its coordinates.

   $\boldsymbol{\left(2\frac{1}{2}, 2\frac{1}{2}\right)}$

   There are infinite correct answers to this question. I could name coordinates that are not on the grid lines. For example, $(1.88, 1.88)$ would be correct.

   h. Name the point where the two axes intersect. Write the coordinates for this point.

   $\boldsymbol{A\ (0, 0)}$

   This point is also known as the origin. The axes meet at the origin.

i. What is the distance between points W and G, or WG?

$\frac{3}{4}$ unit

I count the units between the points. The distance between each grid line is $\frac{1}{4}$.

j. Is the length of $\overline{HG}$ greater than, less than, or equal to $CG + KJ$?

$HG = 2\frac{1}{2}$ units $CG = 1\frac{1}{4}$ units $KJ = 1\frac{1}{2}$ units $CG + KJ = 2\frac{3}{4}$ units $HG < CG + KJ$

k. Janice described how to plot points on the coordinate plane. She said, "If you want to plot $(1,3)$, go 1, and then go 3. Put a point where these lines intersect." Is Janice correct?

Janice is not correct. She should give a starting point and a direction. She should say, "Start at the origin. Along the x-axis, go 1 unit to the right, and then go up 3 units parallel to the y-axis."

Name ____________________ Date ____________

1. Use the grid below to complete the following tasks.

   a. Construct a y-axis that passes through points Y and Z.

   b. Construct a perpendicular x-axis that passes through points Z and X.

   c. Label the origin as 0.

   d. The y-coordinate of W is $2\frac{3}{5}$. Label the whole numbers along the y-axis.

   e. The x-coordinate of V is $2\frac{2}{5}$. Label the whole numbers along the x-axis.

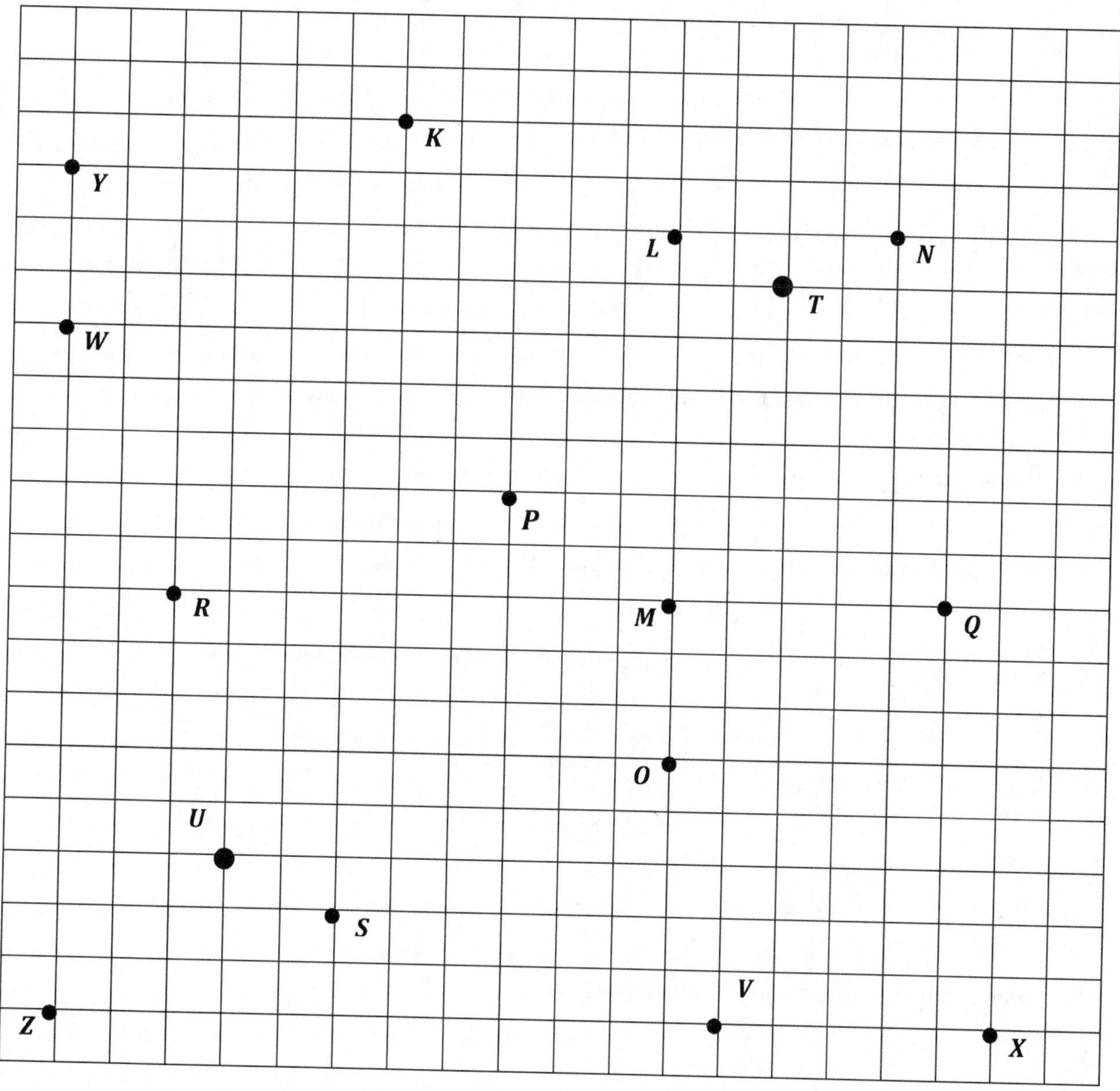

2. For all of the following problems, consider the points K through X on the previous page.

   a. Identify all of the points that have a y-coordinate of $1\frac{3}{5}$.

   b. Identify all of the points that have an x-coordinate of $2\frac{1}{5}$.

   c. Which point is $1\frac{3}{5}$ units above the x-axis *and* $3\frac{1}{5}$ units to the right of the y-axis? Name the point, and give its coordinate pair.

   d. Which point is located $1\frac{1}{5}$ units from the y-axis?

   e. Which point is located $\frac{2}{5}$ unit along the x-axis?

   f. Give the coordinate pair for each of the following points.

   T: ________ U: ________ S: ________ K: ________

   g. Name the points located at the following coordinates.

   $(\frac{3}{5}, \frac{3}{5})$ ________ $(3\frac{2}{5}, 0)$ ________ $(2\frac{1}{5}, 3)$ ________ $(0, 2\frac{3}{5})$ ________

   h. Plot a point whose x- and y-coordinates are equal. Label your point E.

   i. What is the name for the point on the plane where the two axes intersect? ________

   Give the coordinates for this point. (____, ____)

   j. Plot the following points.

   A: $(1\frac{1}{5}, 1)$ B: $(\frac{1}{5}, 3)$ C: $(2\frac{4}{5}, 2\frac{2}{5})$ D: $(1\frac{1}{5}, 0)$

   k. What is the distance between L and N, or LN?

l. What is the distance of MQ?

m. Would RM be greater than, less than, or equal to $LN + MQ$?

n. Leslie was explaining how to plot points on the coordinate plane to a new student, but she left off some important information. Correct her explanation so that it is complete.

"All you have to do is read the coordinates; for example, if it says (4, 7), count four, then seven, and put a point where the two grid lines intersect."

Lesson Notes

The rules for playing *Battleship,* a popular game, are at the end of this Homework Helper.

1. While playing *Battleship,* your friend says, "Hit!" when you guess point (3,2). How do you decide which points to guess next?

   If I get a hit at point $(3, 2)$, then I know I should try to guess one of the four points around $(3, 2)$ because the ship has to lie either vertically or horizontally according to the rules. I would guess one of these points: $(2, 2)$, $(3, 1)$, $(4, 2)$, or $(3, 3)$.

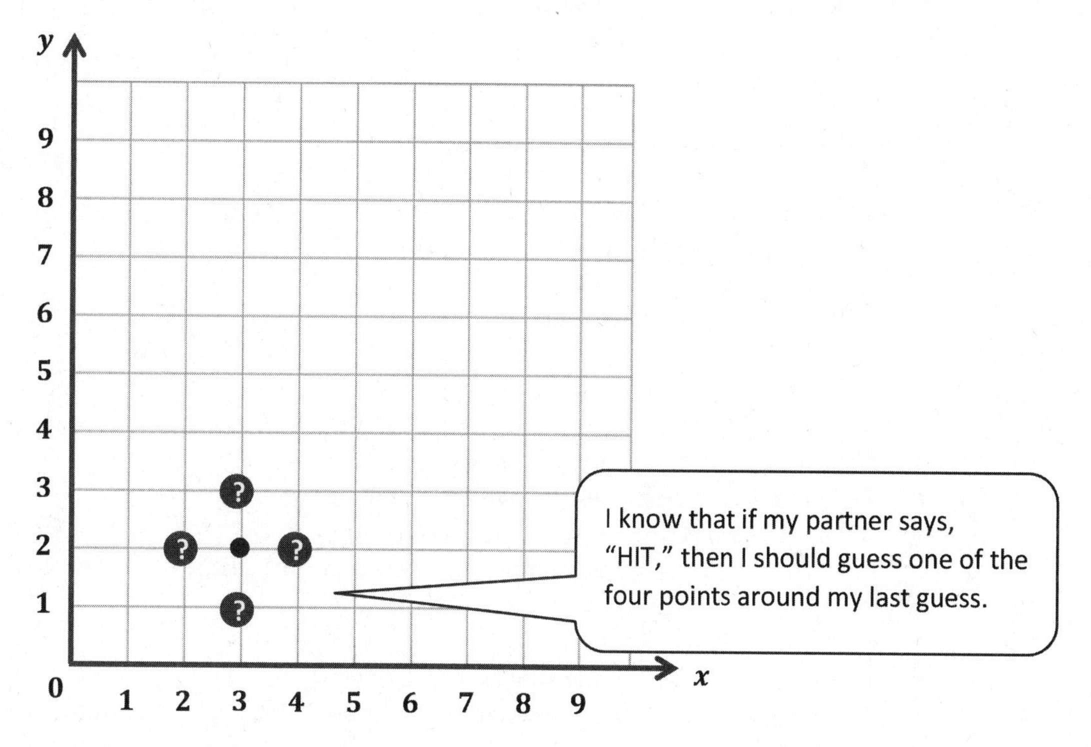

2. What changes to the game could make it more challenging?

   The game is easiest when I count by ones on the coordinate grid's axes. If I changed the axes to count by another number like 7's or 9's on each grid line, the game would be more challenging. It would also be more challenging if I skip-count on the axes by fractions such as $\frac{1}{2}$ or $2\frac{1}{2}$.

Battleship Rules

Goal: To sink all of your opponent's ships by correctly guessing their coordinates.

Materials

- 1 My Ships grid sheet (per person/per game)
- 1 Enemy Ships grid sheet (per person/per game)
- Red crayon/marker for hits
- Black crayon/marker for misses
- Folder to place between players

Ships

- Each player must mark 5 ships on the grid.
  - Aircraft Carrier—Plot 5 points
  - Battleship—Plot 4 points
  - Cruiser—Plot 3 points
  - Submarine—Plot 3 points
  - Patrol Boat—Plot 2 points

Setup

- With your opponent, choose a unit length and fractional unit for the coordinate plane.
- Label chosen units on both grid sheets.
- Secretly select locations for each of the 5 ships on your My Ships grid.
  - All ships must be placed horizontally or vertically on the coordinate plane.
  - Ships can touch each other, but they may not occupy the same coordinate.

Play

- Players take turns firing one shot to attack enemy ships.
- On your turn, call out the coordinates of your attacking shot. Record the coordinates of each attack shot.
- Your opponent checks his My Ships grid. If that coordinate is unoccupied, your opponent says, "Miss." If you named a coordinate occupied by a ship, your opponent says, "Hit."
- Mark each attempted shot on your Enemy Ships grid. Mark a black ✖ on the coordinate if your opponent says, "Miss." Mark a red ✓ on the coordinate if your opponent says, "Hit."
- On your opponent's turn, if he hits one of your ships, mark a red ✓ on that coordinate of your My Ships grid. When one of your ships has every coordinate marked with a ✓, say, "You've sunk my [name of ship]."

Victory

- The first player to sink all (or the most) opposing ships wins.

Name ______________________________ Date ______________

Your homework is to play at least one game of Battleship with a friend or family member. You can use the directions from class to teach your opponent. You and your opponent should record your guesses, hits, and misses on the sheet as you did in class.

When you have finished your game, answer these questions.

1. When you guess a point that is a hit, how do you decide which points to guess next?

2. How could you change the coordinate plane to make the game easier or more challenging?

3. Which strategies worked best for you when playing this game?

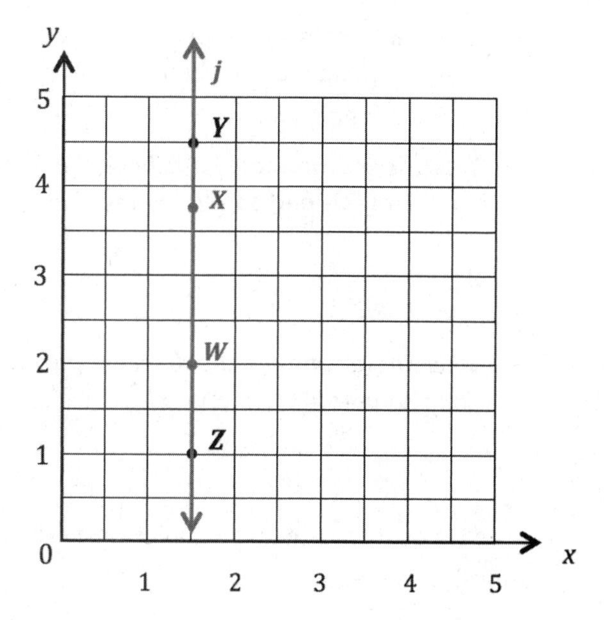

1. Use the coordinate plane to answer the questions.
   a. Use a straight edge to construct a line that goes through points Z and Y. Label this line j.
   b. Line j is perpendicular to the ___*x*___-axis, and is parallel to the ___*y*___-axis.

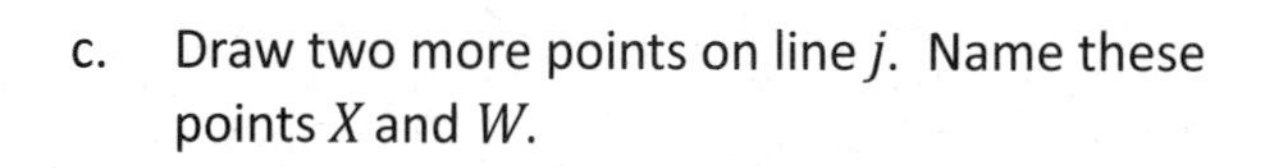

Parallel lines will never cross.

Perpendicular lines form 90° angles.

   c. Draw two more points on line j. Name these points X and W.
   d. Give the coordinates of each point below.

2.
   a. W: $\left(1\frac{1}{2}, 2\right)$ X: $\left(1\frac{1}{2}, 3\frac{3}{4}\right)$ Y: $\left(1\frac{1}{2}, 4\frac{1}{2}\right)$ Z: $\left(1\frac{1}{2}, 1\right)$
   b. What do all these points on line j have in common?

      The x-coordinate is always $\mathbf{1\frac{1}{2}}$.

      This line is perpendicular to the x-axis and parallel to the y-axis because the x-coordinate is the same in every coordinate pair.

   c. Give the coordinate pair of another point that falls on line j with a y-coordinate greater than 10.

      $\left(1\frac{1}{2}, 12\right)$

      As long as the x-coordinate is $1\frac{1}{2}$, the point will fall on line j.

3. For each pair of points below, think about the line that joins them. Will the line be parallel to the x-axis or y-axis? Without plotting them, explain how you know.

   a. $(1.45, 2)$ and $(66, 2)$

   Since these coordinate pairs have the same y-coordinate, the line that joins them will be a horizontal line and parallel to the x-axis.

   b. $\left(\frac{1}{2}, 19\right)$ and $\left(\frac{1}{2}, 82\right)$

   Since these coordinate pairs have the same x-coordinate, the line that joins them will be a vertical line and parallel to the y-axis.

4. Write the coordinate pairs of 3 points that can be connected to construct a line that is $3\frac{1}{8}$ units above and parallel to the x-axis.

   $\left(7, 3\frac{1}{8}\right)$ $\left(6\frac{1}{8}, 3\frac{1}{8}\right)$ $\left(79, 3\frac{1}{8}\right)$

   In order for the line to be $3\frac{1}{8}$ units above the x-axis, the coordinate pairs must have a y-coordinate of $3\frac{1}{8}$. I can use any x-coordinate.

5. Write the coordinate pairs of 3 points that lie on the x-axis.

   $(7, 0)$ $(11.1, 0)$ $(100, 0)$

Name ______________________________ Date ______________

1. Use the coordinate plane to answer the questions.

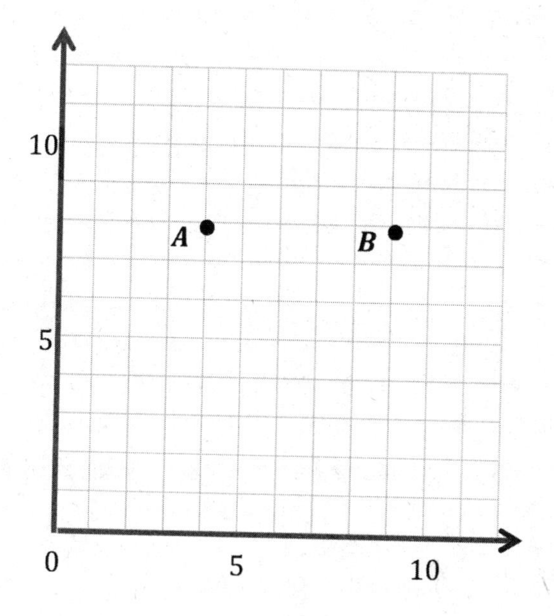

   a. Use a straightedge to construct a line that goes through points A and B. Label this line g.

   b. Line g is parallel to the _______ -axis and is perpendicular to the _______-axis.

   c. Draw two more points on line g. Name them C and D.

   d. Give the coordinates of each point below.

   A: ________ B: ________

   C: ________ D: ________

   e. What do all of the points on line g have in common?

   f. Give the coordinates of another point that falls on line g with an x-coordinate greater than 25.

2. Plot the following points on the coordinate plane to the right.

H: $(\frac{3}{4}, 3)$ I: $(\frac{3}{4}, 2\frac{1}{4})$

J: $(\frac{3}{4}, \frac{1}{2})$ K: $(\frac{3}{4}, 1\frac{3}{4})$

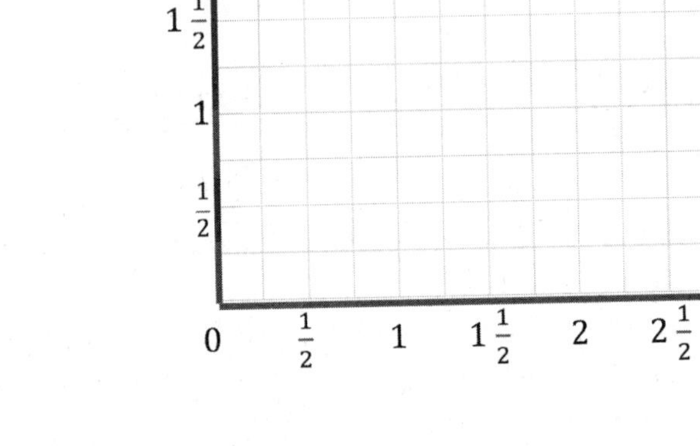

a. Use a straightedge to draw a line to connect these points. Label the line ℓ.

b. In line ℓ, x = ______ for all values of y.

c. Circle the correct word:

Line ℓ is *parallel* *perpendicular* to the x-axis.

Line ℓ is *parallel* *perpendicular* to the y-axis.

d. What pattern occurs in the coordinate pairs that make line ℓ vertical?

3. For each pair of points below, think about the line that joins them. For which pairs is the line parallel to the x-axis? Circle your answer(s). Without plotting them, explain how you know.

a. (3.2, 7) and (5, 7) b. (8, 8.4) and (8, 8.8) c. $(6\frac{1}{2}, 12)$ and (6.2, 11)

4. For each pair of points below, think about the line that joins them. For which pairs is the line parallel to the y-axis? Circle your answer(s). Then, give 2 other coordinate pairs that would also fall on this line.

a. (3.2, 8.5) and (3.22, 24) b. $(13\frac{1}{3}, 4\frac{2}{3})$ and $(13\frac{1}{3}, 7)$ c. (2.9, 5.4) and (7.2, 5.4)

5. Write the coordinate pairs of 3 points that can be connected to construct a line that is $5\frac{1}{2}$ units to the right of and parallel to the y-axis.

   a. ________________ b. ________________ c. ________________

6. Write the coordinate pairs of 3 points that lie on the y-axis.

   a. ________________ b. ________________ c. ________________

7. Leslie and Peggy are playing Battleship on axes labeled in halves. Presented in the table is a record of Peggy's guesses so far. What should she guess next? How do you know? Explain using words and pictures.

(5, 5)	miss
(4, 5)	hit
$(3\frac{1}{2}, 5)$	miss
$(4\frac{1}{2}, 5)$	miss

1. Plot and label the following points on the coordinate plane.

   $K\ (0.7, 0.6)$ $P\ (0.7, 1.1)$ $M\ (0.2, 0.3)$ $H\ (0.9, 0.3)$

   a. Use a straightedge to construct line segments KP and MH.

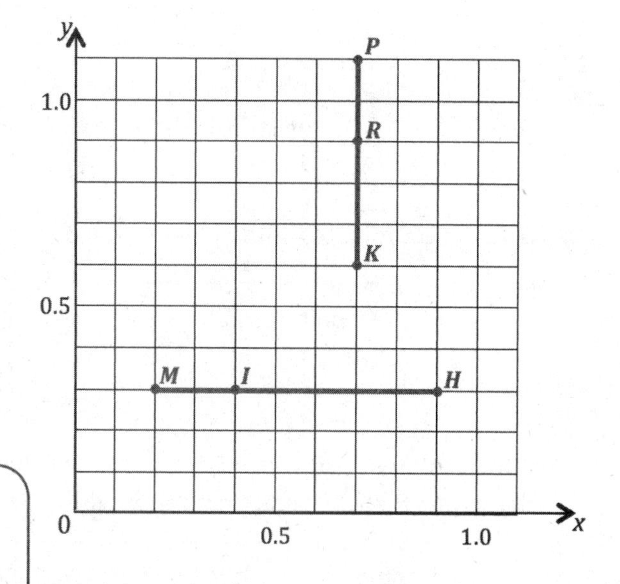

   b. Name the line segment that is perpendicular to the x-axis and parallel to the y-axis.

   $\overline{KP}$

   > Because the x-coordinates of K and P are the same, segment KP is parallel to the y-axis.

   c. Name the line segment that is parallel to the x-axis and perpendicular to the y-axis.

   $\overline{MH}$

   > Because the y-coordinates of M and H are the same, segment MH is perpendicular to the y-axis.

   d. Plot a point on $\overline{KP}$, and name it R.

   e. Plot a point on $\overline{MH}$, and name it I.

   f. Write the coordinates for points R and I.

   $R\ (0.7, 0.9)$ $I\ (0.4, 0.3)$

2. Construct line j such that the y-coordinate of every point is $2\frac{1}{4}$, and construct line k such that the x-coordinate of every point is $1\frac{3}{4}$.

Since all the y-coordinates are the same, line j will be a horizontal line. Since all the x-coordinates are the same, line k will be a vertical line.

a. Line j is $\mathbf{2\frac{1}{4}}$ units from the x-axis.

b. Give the coordinates of the point on line j that is 1 unit from the y-axis.

$\left(\mathbf{1, 2\frac{1}{4}}\right)$

"1 unit from the y-axis" gives the value of the x-coordinate.

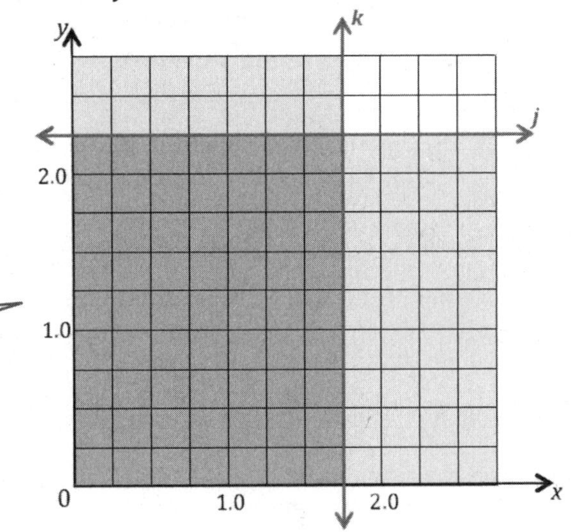

c. With a colored pencil, shade the portion of the grid that is less than $2\frac{1}{4}$ units from the x-axis.

I use blue to shade the grid below line j.

d. Line k is $\mathbf{1\frac{3}{4}}$ units from the y-axis.

e. Give the coordinates of the point on line k that is $1\frac{1}{2}$ units from the x-axis.

$\left(\mathbf{1\frac{3}{4}, 1\frac{1}{2}}\right)$

"$1\frac{1}{2}$ units from the x-axis" gives the value of the y-coordinate.

f. With another colored pencil, shade the portion of the grid that is less than $1\frac{3}{4}$ units from the y-axis.

I use pink to shade the grid to the left of line k. The area of the grid that is below line j and to the left of line k now looks purple.

Name ______________________________ Date ______________

1. Plot and label the following points on the coordinate plane.

C: (0.4, 0.4) A: (1.1, 0.4) S: (0.9, 0.5) T: (0.9, 1.1)

a. Use a straightedge to construct line segments $\overline{CA}$ and $\overline{ST}$.

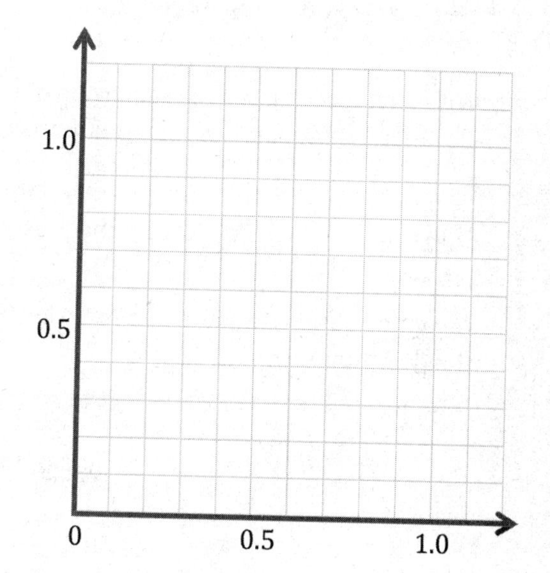

b. Name the line segment that is perpendicular to the x-axis and parallel to the y-axis.

____

c. Name the line segment that is parallel to the x-axis and perpendicular to the y-axis.

____

d. Plot a point on $\overline{CA}$, and name it E. Plot a point on line segment $\overline{ST}$, and name it R.

e. Write the coordinates of points E and R.

E (____, ____) R (____, ____)

2. Construct line m such that the y-coordinate of every point is $1\frac{1}{2}$, and construct line n such that the x-coordinate of every point is $5\frac{1}{2}$.

   a. Line m is ________ units from the x-axis.

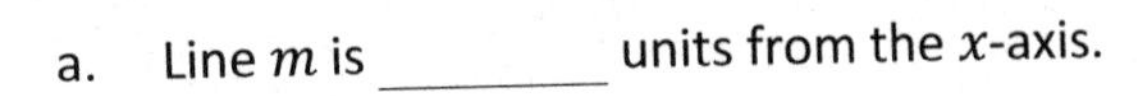

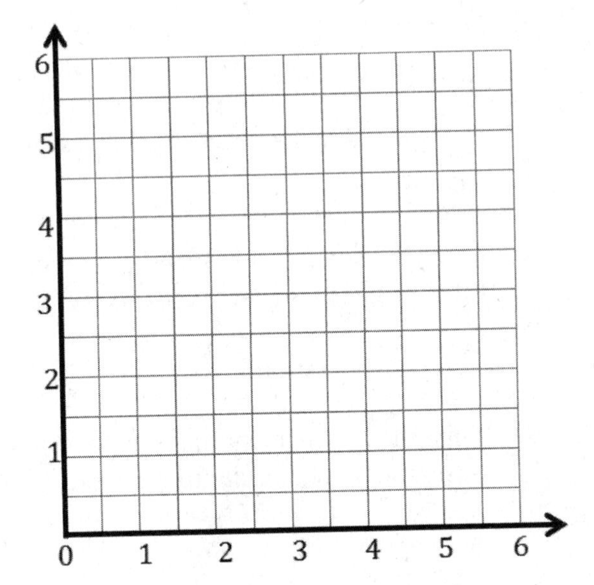

   b. Give the coordinates of the point on line m that is 2 units from the y-axis. ________

   c. With a blue pencil, shade the portion of the grid that is less than $1\frac{1}{2}$ units from the x-axis.

   d. Line n is ________ units from the y-axis.

   e. Give the coordinates of the point on line n that is $3\frac{1}{2}$ units from the x-axis.

   f. With a red pencil, shade the portion of the grid that is less than $5\frac{1}{2}$ units from the y-axis.

3. Construct and label lines e, r, s, and o on the plane below.

    a. Line e is 3.75 units above the x-axis.

    b. Line r is 2.5 units from the y-axis.

    c. Line s is parallel to line e but 0.75 farther from the x-axis.

    d. Line o is perpendicular to lines s and e and passes through the point $(3\frac{1}{4}, 3\frac{1}{4})$.

4. Complete the following tasks on the plane.

    a. Using a blue pencil, shade the region that contains points that are more than $2\frac{1}{2}$ units and less than $3\frac{1}{4}$ units from the y-axis.

    b. Using a red pencil, shade the region that contains points that are more than $3\frac{3}{4}$ units and less than $4\frac{1}{2}$ units from the x-axis.

    c. Plot a point that lies in the double-shaded region, and label its coordinates.

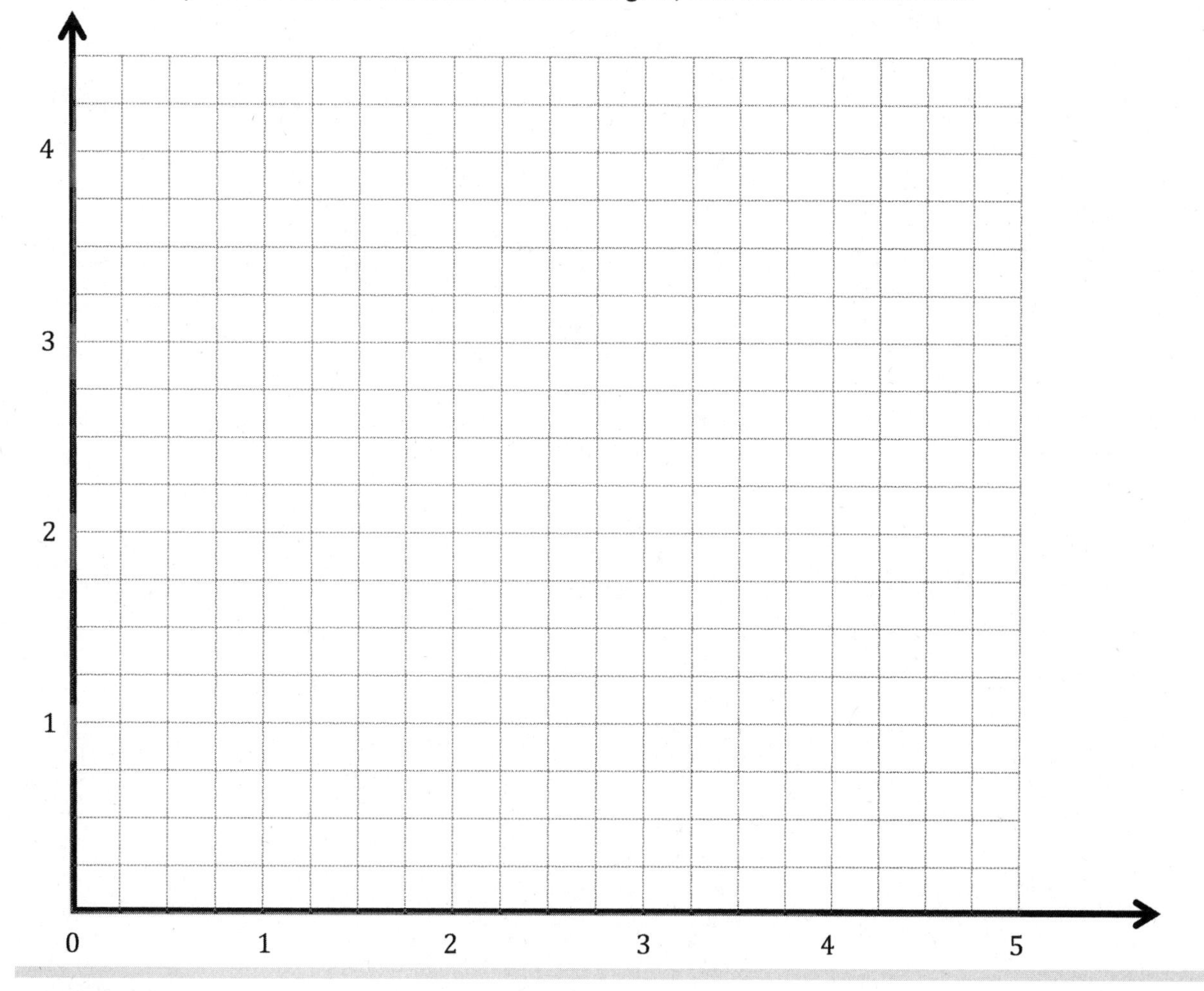

1. Complete the chart. Then, plot the points on the coordinate plane.

x	y	(x, y)
3	$1\frac{1}{2}$	$\left(3, 1\frac{1}{2}\right)$
$1\frac{1}{2}$	0	$\left(1\frac{1}{2}, 0\right)$
2	$\frac{1}{2}$	$\left(2, \frac{1}{2}\right)$
$4\frac{1}{2}$	3	$\left(4\frac{1}{2}, 3\right)$

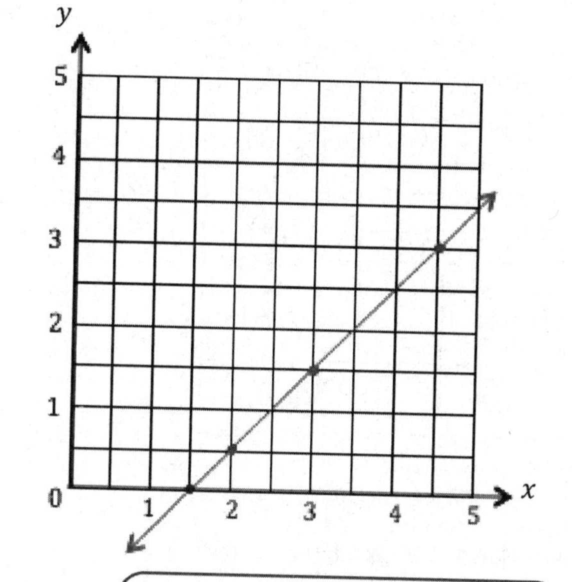

a. Use a straightedge to draw a line connecting these points.

b. Write a rule showing the relationship between the x- coordinates and y-coordinates of points on this line.

I could have also said that the y-coordinates are $1\frac{1}{2}$ less than the corresponding x-coordinates.

Each x-coordinate is $1\frac{1}{2}$ more than its corresponding y-coordinate.

c. Name the coordinates of two other points that are also on this line.

$\left(2\frac{1}{2}, 1\right)$ *and* $\left(5, 3\frac{1}{2}\right)$

As long as the x-coordinate is $1\frac{1}{2}$ more than the y-coordinate, the point will fall on this line.

2. Complete the chart. Then, plot the points on the coordinate plane.

x	y	(x, y)
$\frac{3}{4}$	3	$\left(\frac{3}{4}, 3\right)$
1	4	$(1, 4)$
$\frac{1}{2}$	2	$\left(\frac{1}{2}, 2\right)$
0	0	$(0, 0)$

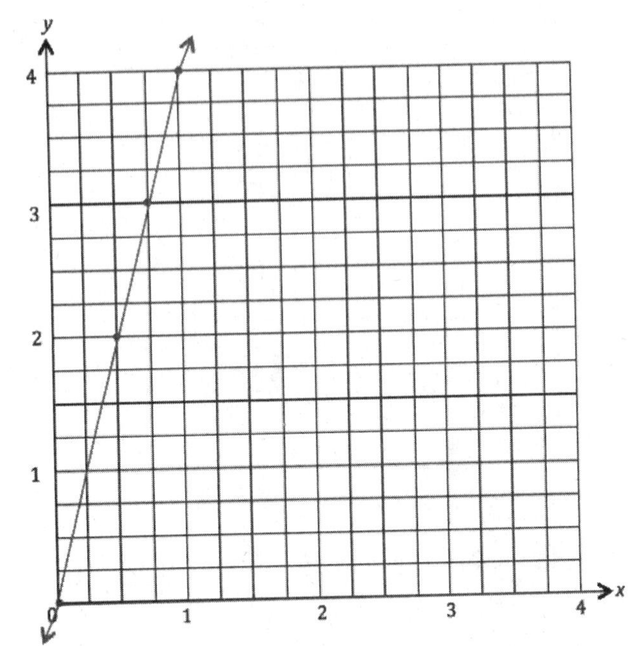

a. Use a straightedge to draw a line connecting these points.

b. Write a rule showing the relationship between the x-coordinates and y-coordinates for points on the line.

Each y-coordinate is four times as much as its corresponding x-coordinate.

c. Name two other points that are also on this line.

$(2, 8)$ ***and*** $\left(\frac{5}{8}, 2\frac{1}{2}\right)$

This rule is also correct: Each x-coordinate is 1 fourth as much as its corresponding y-coordinate.

3. Use the coordinate plane to answer the following questions.

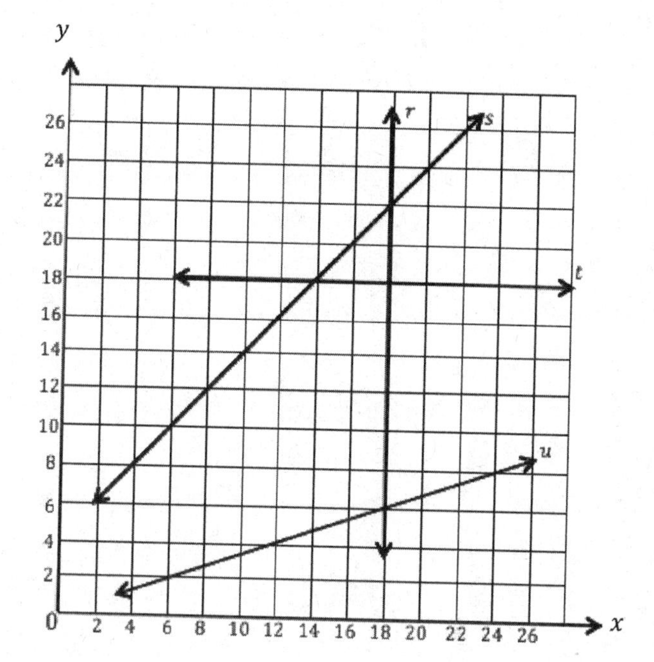

a. For any point on line r, the x-coordinate is **18**.

The x-coordinate tells the distance from the y-axis.

b. Give the coordinates for 3 points that are on line s.

(4, 8) (10, 14) (20, 24)

c. Write a rule that describes the relationship between the x-coordinates and y-coordinates on line s.

Each y-coordinate is 4 more than its corresponding x-coordinate.

I could also say, "Each x-coordinate is 4 less than the y-coordinate."

d. Give the coordinates for 3 points that are on line u.

(6, 2) (12, 4) (24, 8)

e. Write a rule that describes the relationship between the x-coordinates and y-coordinates on line u.

Each x-coordinate is 3 times as much as the y-coordinate.

I could also say, "Each y-coordinate is $\frac{1}{3}$ the value of the x-coordinate."

f. Each of these points lies on at least 1 of the lines shown in the plane above. Identify a line that contains the following points.

$(18, 16.3)$ **r** $(9.5, 13.5)$ **s** $\left(16, 5\frac{1}{3}\right)$ **u** $(22.3, 18)$ **t**

All of the points on line r have an x-coordinate of 18.

All of the points on line t have a y-coordinate of 18.

Name ____________________ Date ____________

1. Complete the chart. Then, plot the points on the coordinate plane.

x	y	(x, y)
2	0	
$3\frac{1}{2}$	$1\frac{1}{2}$	
$4\frac{1}{2}$	$2\frac{1}{2}$	
6	4	

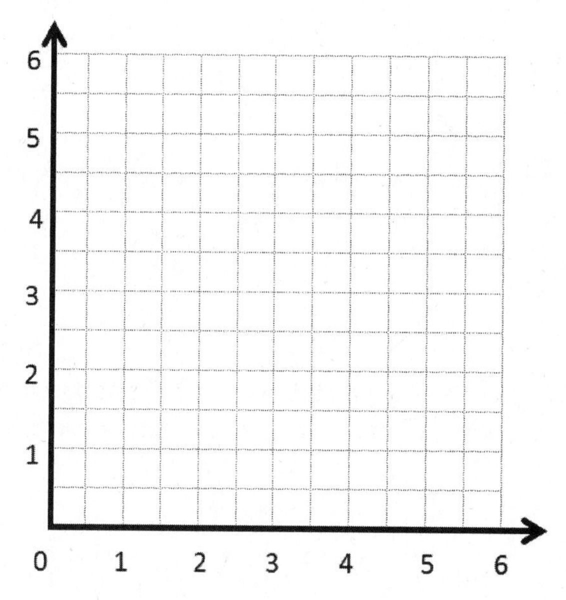

a. Use a straightedge to draw a line connecting these points.

b. Write a rule showing the relationship between the x- and y-coordinates of points on this line.

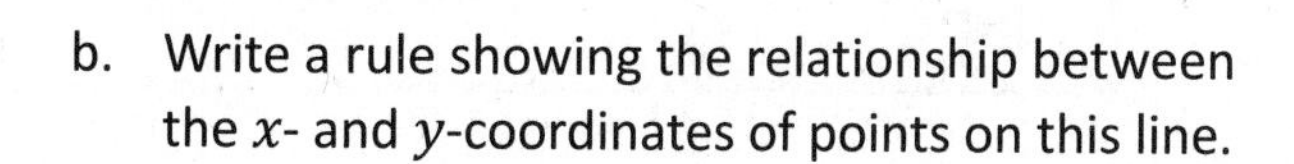

c. Name two other points that are also on this line. __________ __________

2. Complete the chart. Then, plot the points on the coordinate plane.

x	y	(x, y)
0	0	
$\frac{1}{4}$	$\frac{3}{4}$	
$\frac{1}{2}$	$1\frac{1}{2}$	
1	3	

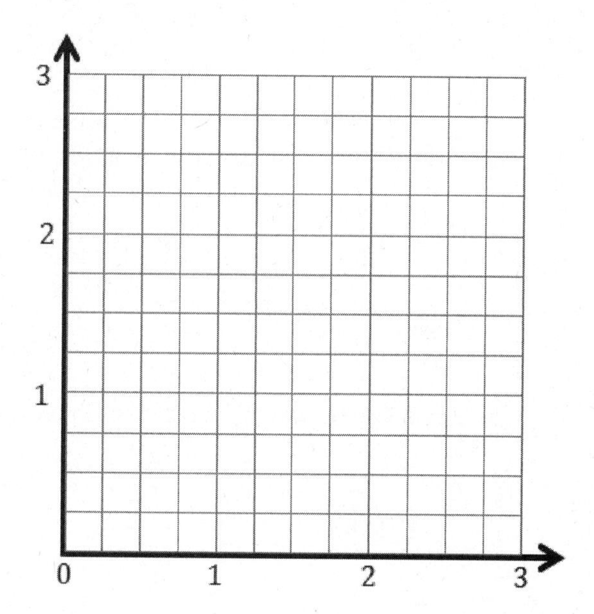

a. Use a straightedge to draw a line connecting these points.

b. Write a rule showing the relationship between the x- and y-coordinates for points on the line.

c. Name two other points that are also on this line. __________ __________

3. Use the coordinate plane to answer the following questions.

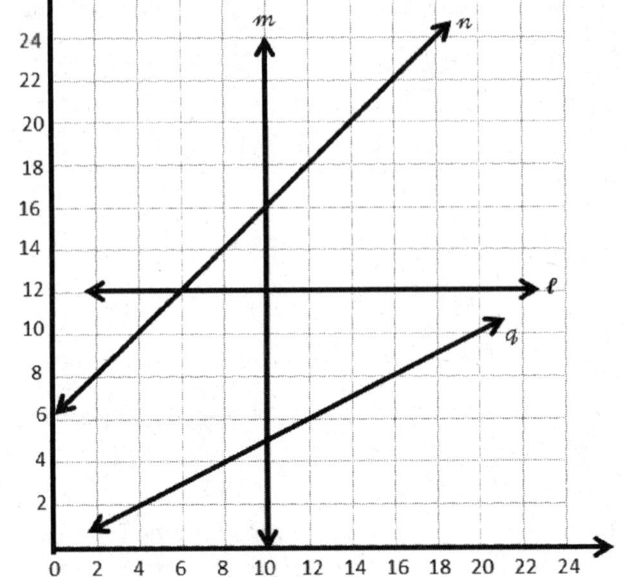

a. For any point on line m, the x-coordinate is ________.

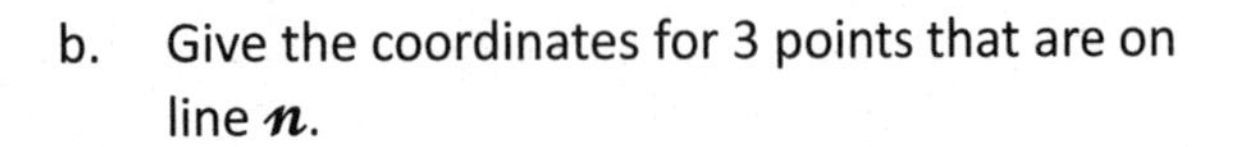

b. Give the coordinates for 3 points that are on line n.

c. Write a rule that describes the relationship between the x- and y-coordinates on line n.

d. Give the coordinates for 3 points that are on line q.

e. Write a rule that describes the relationship between the x- and y-coordinates on line q.

f. Identify a line on which each of these points lie.

i. (10, 3.2) ______ ii. (12.4, 18.4) ______

iii. (6.45, 12) ______ iv. (14, 7) ______

Complete this table such that each y-coordinate is 5 more than the corresponding x-coordinate.

x	y	(x, y)
2	7	$(2, 7)$
4	9	$(4, 9)$
6	11	$(6, 11)$

I choose coordinate pairs that satisfy the rule and will fit on the coordinate plane.

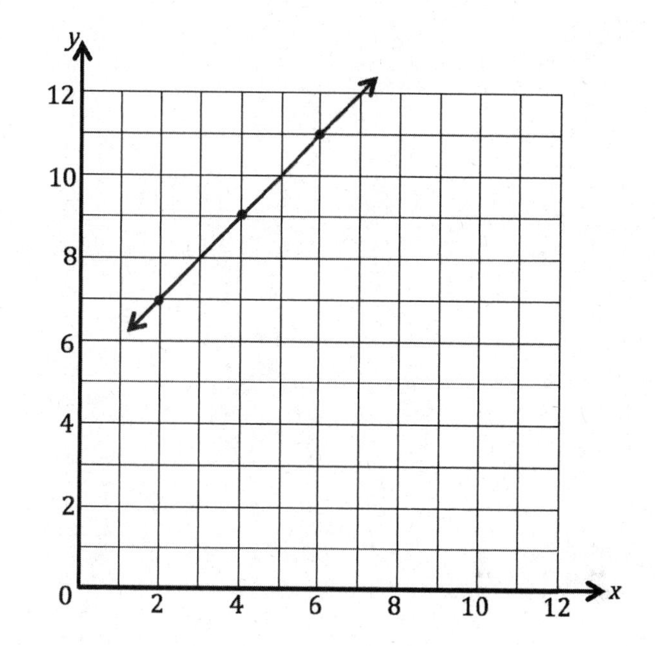

a. Plot each point on the coordinate plane.

b. Use a straightedge to construct a line connecting these points.

c. Give the coordinate of 3 other points that fall on this line with x-coordinates greater than 15.

$(17, 22)$ $\left(20\frac{1}{2}, 25\frac{1}{2}\right)$ $(100, 105)$

Although I can't see these points on the plane, I know they will fall on the line because each y-coordinate is 5 more than the x-coordinate.

Name ________________________________ Date ________________

1. Complete this table such that each y-coordinate is 4 more than the corresponding x-coordinate.

x	y	(x, y)

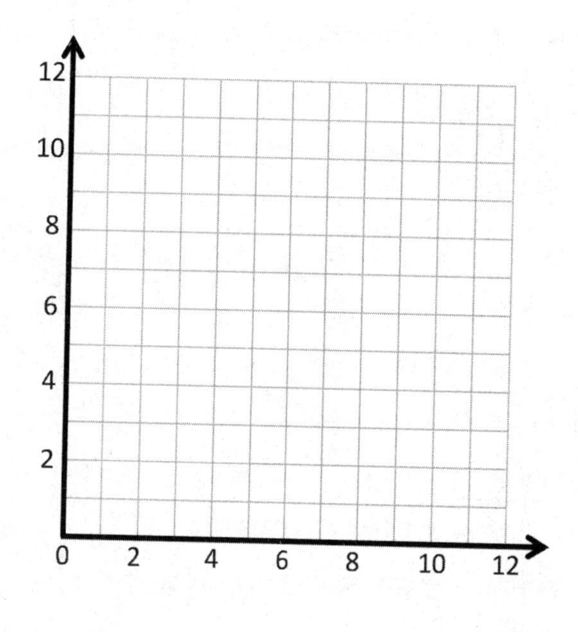

a. Plot each point on the coordinate plane.

b. Use a straightedge to construct a line connecting these points.

c. Give the coordinates of 2 other points that fall on this line with x-coordinates greater than 18.
(______, ______) and (______, ______)

2. Complete this table such that each y-coordinate is 2 times as much as its corresponding x-coordinate.

x	y	(x, y)

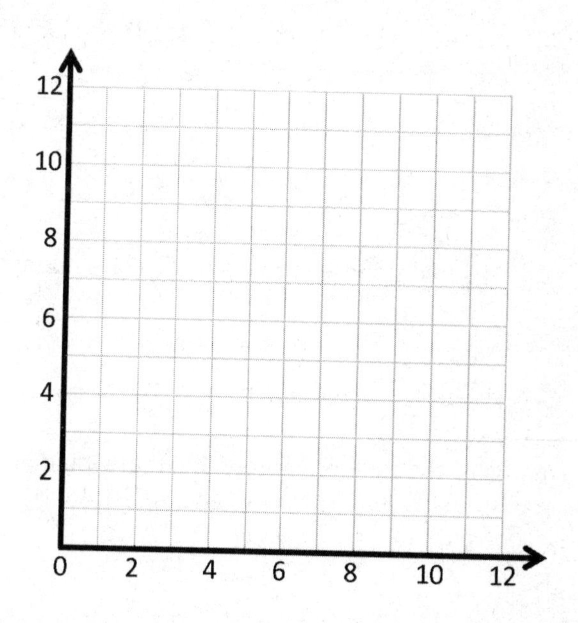

a. Plot each point on the coordinate plane.

b. Use a straightedge to draw a line connecting these points.

c. Give the coordinates of 2 other points that fall on this line with y-coordinates greater than 25.
(______, ______) and (______, ______)

3. Use the coordinate plane below to complete the following tasks.

a. Graph these lines on the plane.

line ℓ: x is equal to y

	x	y	(x, y)
A			
B			
C			

line m: y is 1 less than x

	x	y	(x, y)
G			
H			
I			

line n: y is 1 less than twice x

	x	y	(x, y)
S			
T			
U			

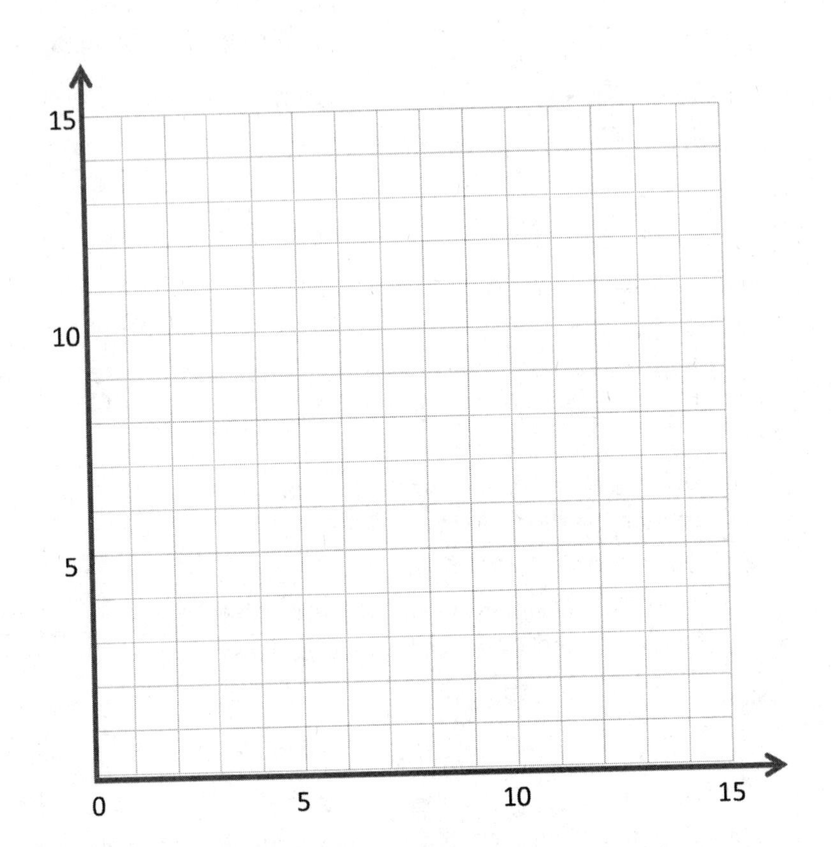

b. Do any of these lines intersect? If yes, identify which ones, and give the coordinates of their intersection.

c. Are any of these lines parallel? If yes, identify which ones.

d. Give the rule for another line that would be parallel to the lines you listed in Problem 3(c).

In order to find the y-coordinates, I just follow the rule, "y is 2 less than x."

So when x is 5, I find the number that is 2 less than 5. $5 - 2 = 3$.

So when x is 5, y is 3.

1. Complete the table with the given rules.

Line a

Rule: y *is* 2 *less than* x.

x	y	(x, y)
2	**0**	**(2, 0)**
5	**3**	**(5, 3)**
10	**8**	**(10, 8)**
17	**15**	**(17, 15)**

Line b

Rule: y *is* 4 *less than* x.

x	y	(x, y)
5	**1**	**(5, 1)**
8	**4**	**(8, 4)**
14	**10**	**(14, 10)**
20	**16**	**(20, 16)**

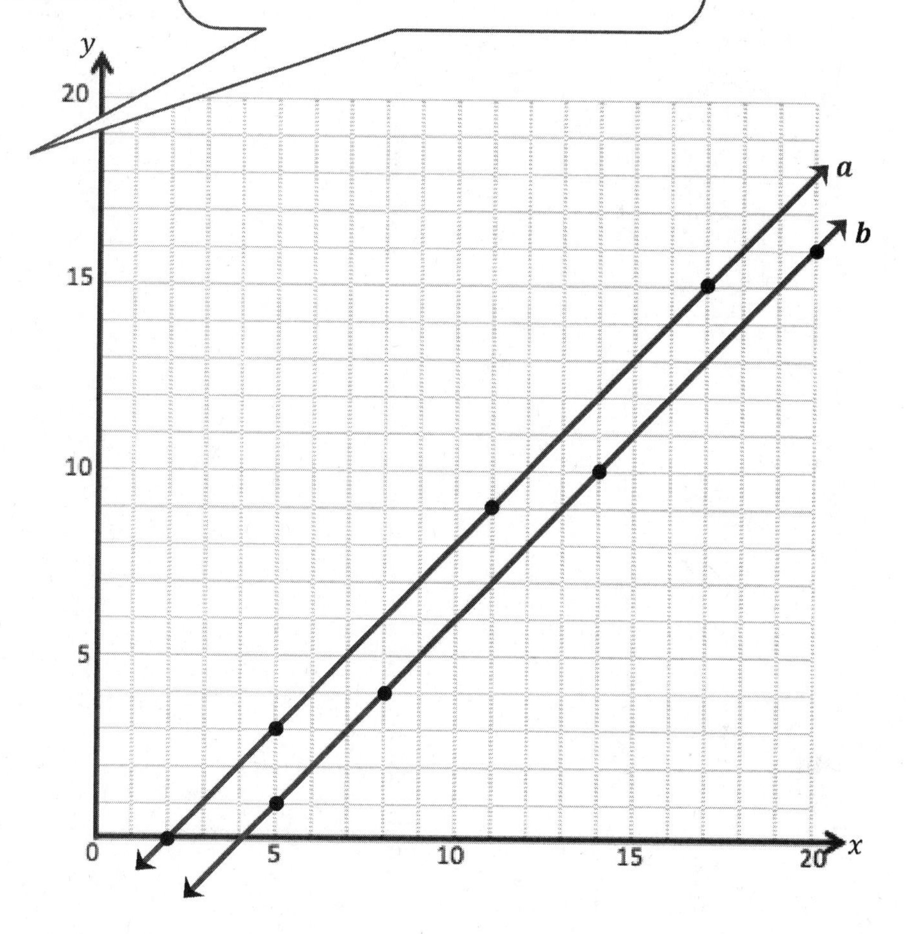

a. Construct each line on the coordinate plane.

b. Compare and contrast these lines.

The lines are parallel. Neither line passes through the origin. Line b looks like it is closer to the x-axis or farther down and to the right. Line a is closer to the y-axis and farther up and to the left

c. Based on the patterns you see, predict what line c, whose rule is y *is* 6 *less than* x, would look like.

Since the rule for line c is also a subtraction rule, I think it will also be parallel to lines a and b. But, since the rule is "y is 6 less than x," I think it will be even farther to the right than line b.

2. Complete the table for the given rules.

In order to find the y-coordinates, I just follow the rule, "y is 2 times as much as x."

So when x is 4, I find the number that is 2 times as much as 4: $4 \times 2 = 8$.

So when x is 4, y is 8.

Line e

Rule: y is 2 times as much as x.

x	y	(x, y)
0	**0**	**(0, 0)**
1	**2**	**(1, 2)**
4	**8**	**(4, 8)**
9	**18**	**(9, 18)**

Line f

Rule: y is half as much as x.

x	y	(x, y)
0	**0**	**(0, 0)**
6	**3**	**(6, 3)**
12	**6**	**(12, 6)**
18	**9**	**(18, 9)**

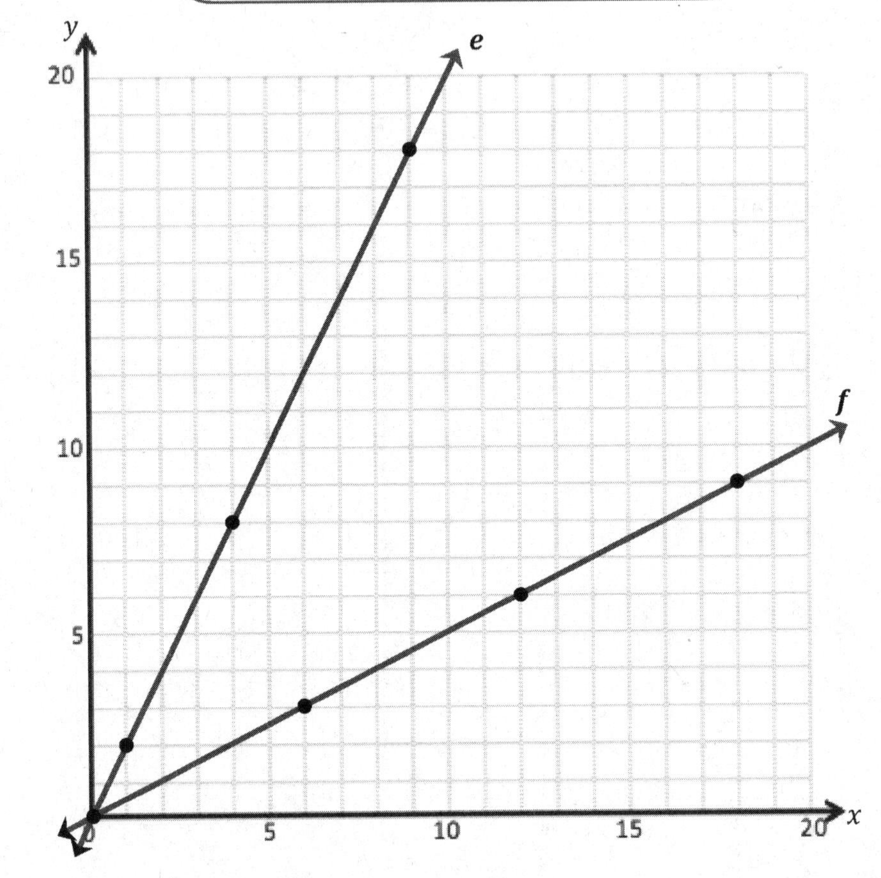

a. Construct each line on the coordinate plane.

b. Compare and contrast these lines.

Both lines go through the origin, and they are not parallel. Line e is steeper than line f.

c. Based on the patterns you see, predict what line g, whose rule is *y is 3 times as much as x*, and line h, whose rule is *y is a third as much as x*, would look like.

Since the rule for line g is also a multiplication rule, I think it will also pass through the origin. But, since the rule is "y is 3 times as much as x," I think it will be even steeper than lines e and f.

Name ____________________ Date ____________

1. Complete the table for the given rules.

Line a

Rule: y is 1 less than x

x	y	(x, y)
1		
4		
9		
16		

Line b

Rule: y is 5 less than x

x	y	(x, y)
5		
8		
14		
20		

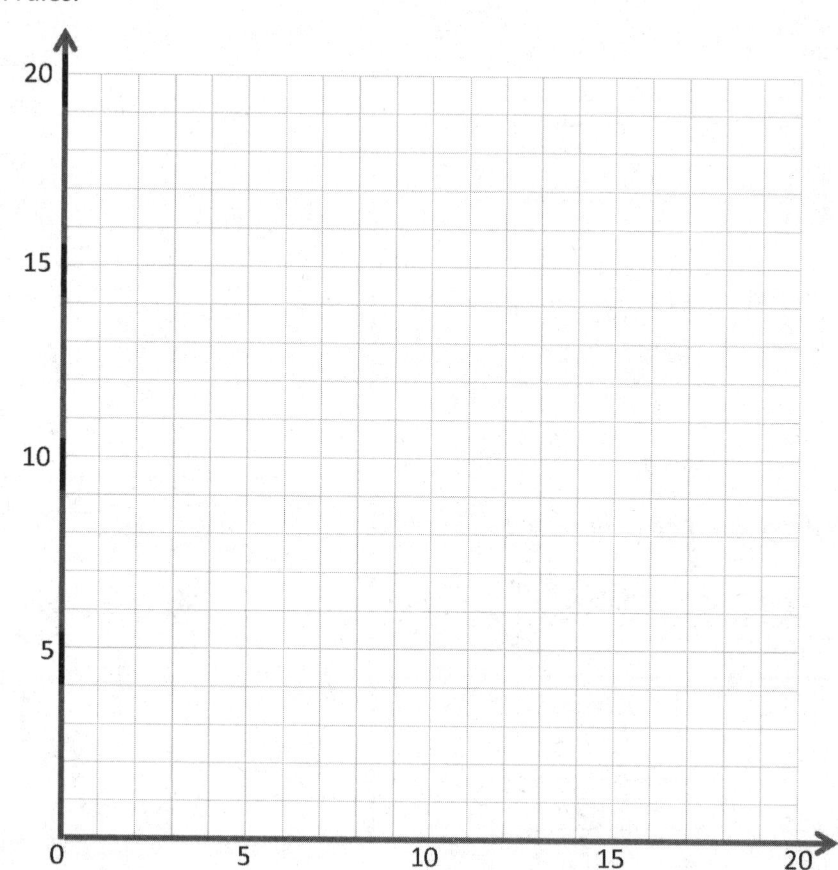

a. Construct each line on the coordinate plane.

b. Compare and contrast these lines.

c. Based on the patterns you see, predict what line c, whose rule is y is 7 less than x, would look like. Draw your prediction on the plane above.

2. Complete the table for the given rules.

Line e

Rule: y is 3 times as much as x

x	y	(x, y)
0		
1		
4		
6		

Line f

Rule: y is a third as much as x

x	y	(x, y)
0		
3		
9		
15		

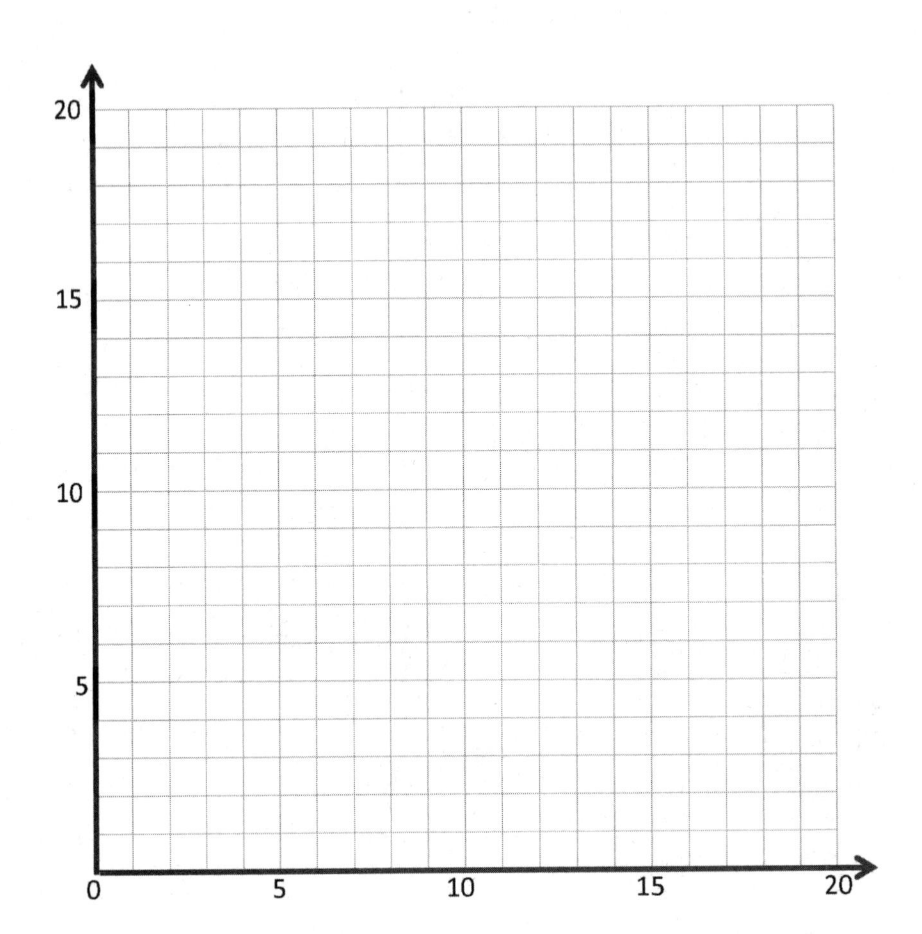

a. Construct each line on the coordinate plane.

b. Compare and contrast these lines.

c. Based on the patterns you see, predict what line g, whose rule is *y is 4 times as much as x*, and line h, whose rule is *y is one-fourth as much as x*, would look like. Draw your prediction in the plane above.

1. Use the coordinate plane to complete the following tasks.

   a. The rule for line b is "x and y are equal." Construct line b.

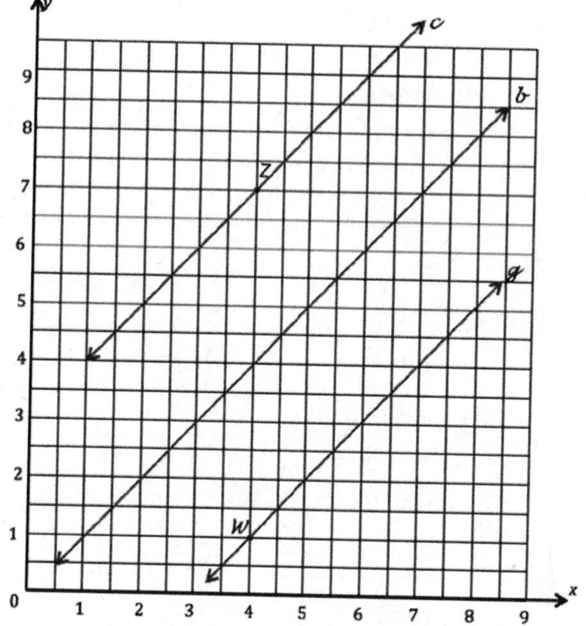

Some coordinate pairs that follow this rule are
$(1, 1)$ $(3, 3)$ $(6.5, 6.5)$

   b. Construct a line, c, that is parallel to line b and contains point Z.

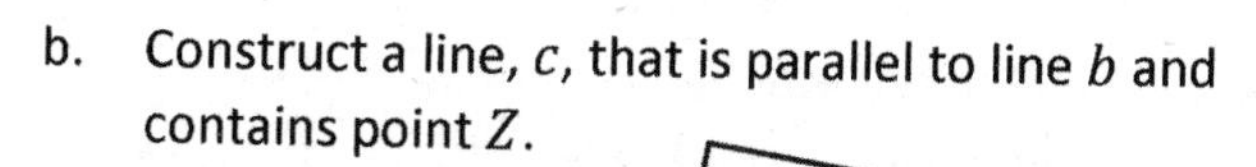

Since line c needs to be parallel to line b, the rule for line c must be an addition or subtraction rule. The coordinate pair for Z is $(4, 7)$, so I can draw line c along other coordinate pairs that have a y-coordinate that is 3 *more* than the x-coordinate.

   c. Name 3 coordinate pairs on line c.

   (2, 5) **(3, 6)** **(6, 9)**

   d. Identify a rule to describe line c.

   x is 3 less than y.

Another way to describe this rule is: y is 3 more than x.

   e. Construct a line, g, that is parallel to line b and contains point W.

   f. Name 3 points on line g.

   (3.5, 0.5) **(6, 3)** **(7, 4)**

Again, since line g needs to be parallel to line b, the rule for line g must be an addition or subtraction rule. The coordinate pair for W is $(4, 1)$, so I can draw line g along other coordinate pairs that have a y-coordinate that is 3 *less* than the x-coordinate.

   g. Identify a rule to describe line g.

   x is 3 more than y.

h. Compare and contrast lines c and g in terms of their relationship to line b.

Lines c and g are both parallel to line b.
Line c is above line b because the points on line c have y-coordinates greater than the x-coordinates.
Line g is below line b because the points on line g have y-coordinates less than the x-coordinates.

2. Write a rule for a fourth line that would be parallel to those in Problem 1 and that would contain the point $(5, 6)$.

y is 1 more than x.

Because this line is parallel to the others, I know it has to be an addition rule. In the given coordinate pair, the y-coordinate is 1 more than the x-coordinate.

3. Use the coordinate plane below to complete the following tasks.

a. Line b represents the rule "x and y are equal."

I can also think of this as a multiplication rule. "x times 1 is equal to y."

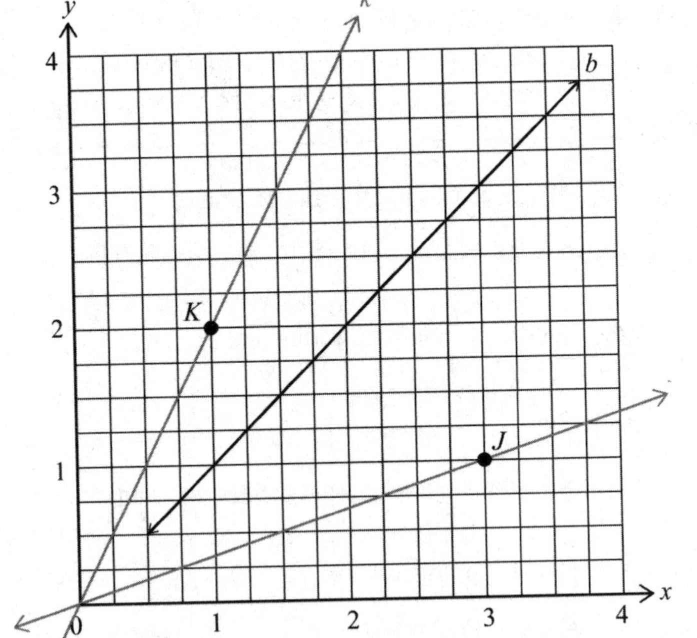

b. Construct a line, j, that contains the origin and point J.

c. Name 3 points on line j.

$\mathbf{(3, 1)}$ $\mathbf{\left(1\frac{1}{2}, \frac{1}{2}\right)}$ $\mathbf{\left(\frac{3}{4}, \frac{1}{4}\right)}$

d. Identify a rule to describe line j.

x is 3 times as much as y.

As I analyze the relationship between the x- and y-coordinates on line j, I can see that each y-coordinate is $\frac{1}{3}$ the value of its corresponding x-coordinate.

e. Construct a line, k, that contains the origin and point K.

f. Name 3 points on line k.

$\left(\frac{1}{2}, 1\right)$ $\left(1\frac{1}{2}, 3\right)$ $(2, 4)$

g. Identify a rule to describe line k.

x is half of y.

As I analyze the relationship between the x-coordinates and y-coordinates on line k, I can see that each y-coordinate is twice the value of its corresponding x-coordinate.

Name ______________________________ Date ______________

1. Use the coordinate plane to complete the following tasks.

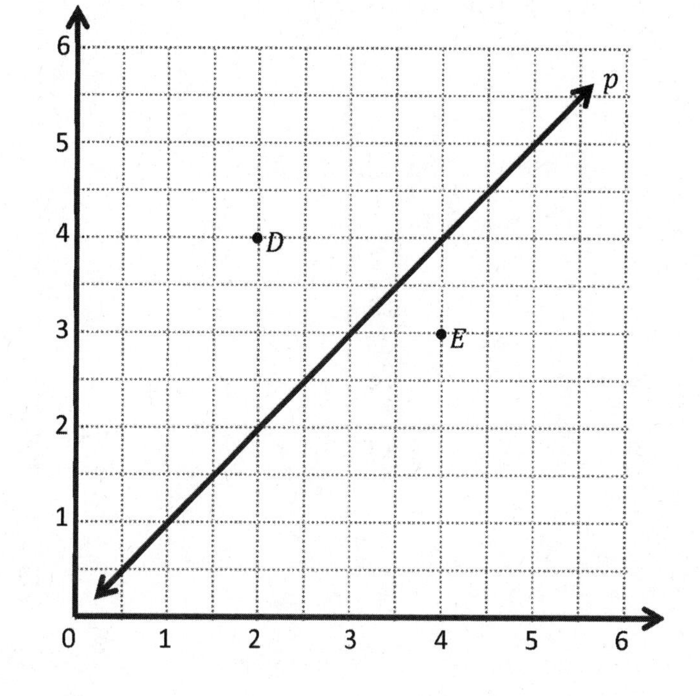

   a. Line p represents the rule x *and* y *are equal*.

   b. Construct a line, d, that is parallel to line p and contains point D.

   c. Name 3 coordinate pairs on line d.

   d. Identify a rule to describe line d.

   e. Construct a line, e, that is parallel to line p and contains point E.

   f. Name 3 points on line e.

   g. Identify a rule to describe line e.

   h. Compare and contrast lines d and e in terms of their relationship to line p.

2. Write a rule for a fourth line that would be parallel to those above and that would contain the point $(5\frac{1}{2}, 2)$. Explain how you know.

3. Use the coordinate plane below to complete the following tasks.

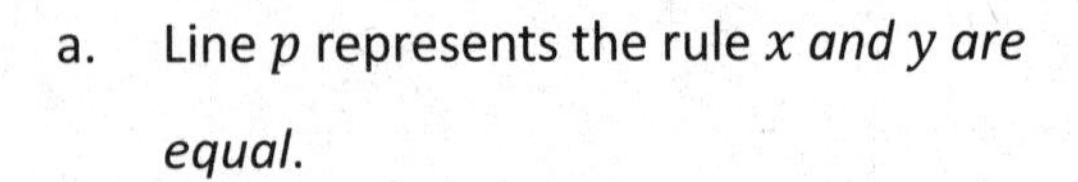

a. Line p represents the rule x *and* y *are equal*.

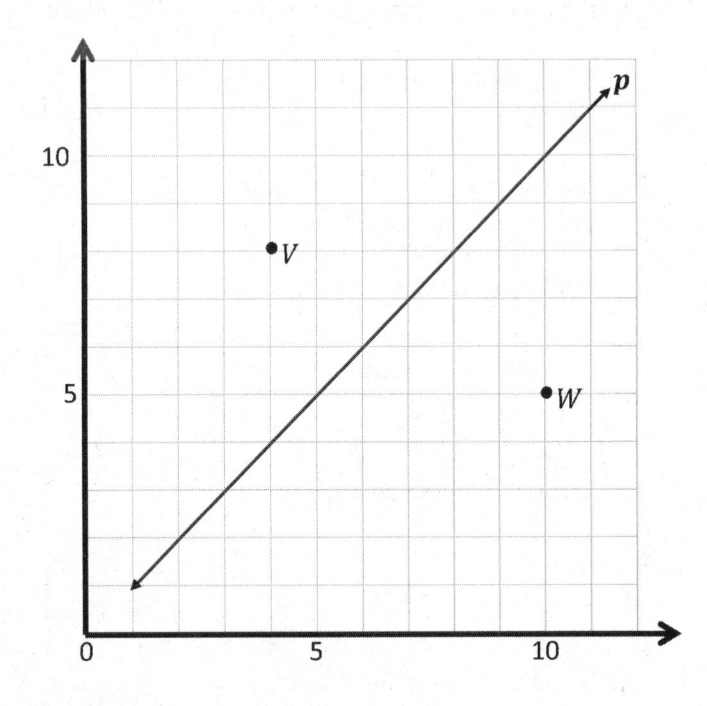

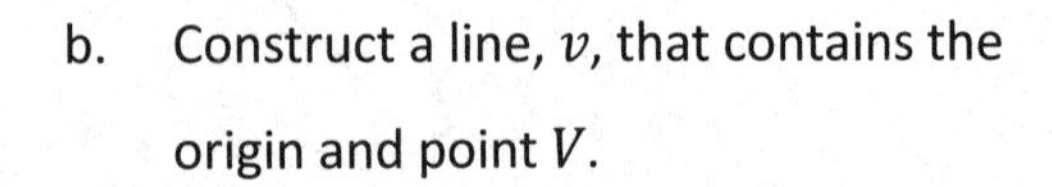

b. Construct a line, v, that contains the origin and point V.

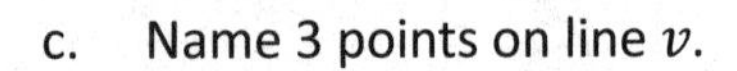

c. Name 3 points on line v.

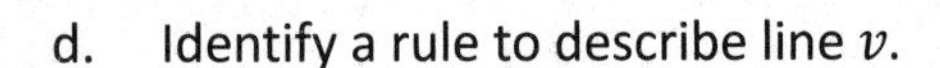

d. Identify a rule to describe line v.

e. Construct a line, w, that contains the origin and point W.

f. Name 3 points on line w.

g. Identify a rule to describe line w.

h. Compare and contrast lines v and w in terms of their relationship to line p.

i. What patterns do you see in lines that are generated by multiplication rules?

1. Complete the tables for the given rules.

Line p

Rule: *Halve* x.

x	y	(x, y)
2	**1**	**(2, 1)**
4	**2**	**(4, 2)**
6	**3**	**(6, 3)**

Line q

Rule: *Halve* x, *and then add* 1.

x	y	(x, y)
2	**2**	**(2, 2)**
4	**3**	**(4, 3)**
6	**4**	**(6, 4)**

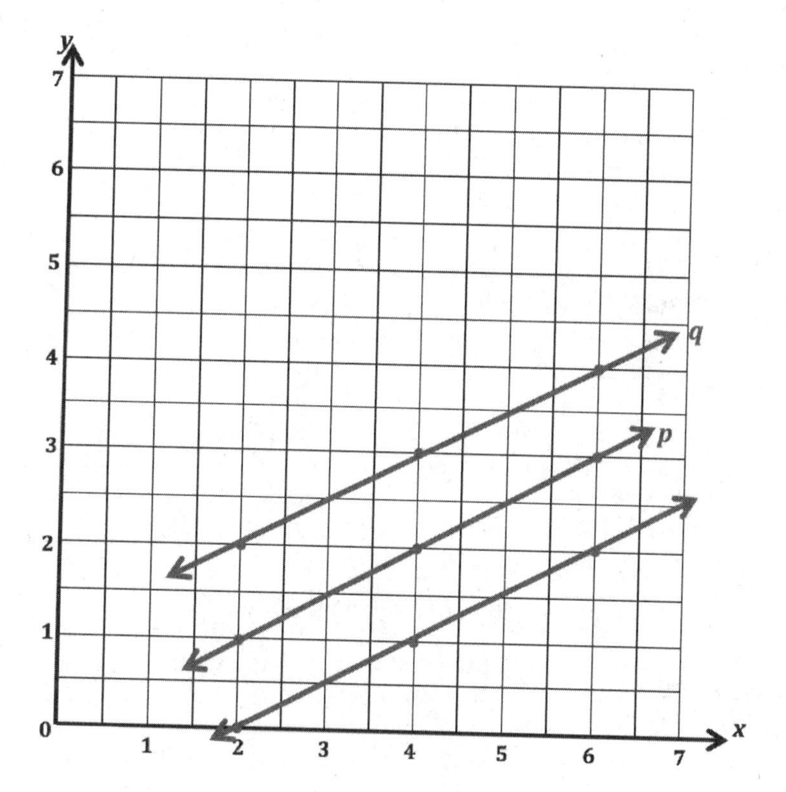

a. Draw each line on the coordinate plane above.

b. Compare and contrast these lines.

Line q is *above* line p because the rule says, "*then add* 1."

They are parallel lines. Line q is above line p. The distance between the two lines is 1 unit.

c. Based on the patterns you see, predict what the line for the rule "halve x, and then subtract 1" would look like. Draw your prediction on the plane above.

I predict the line will be parallel to lines p and q.

It will be 1 unit below line p because the rule says, "then subtract 1."

I need to look for coordinate pairs that follow the rule, "*double x, and then add* $\frac{1}{2}$."

2. Circle the point(s) that the line for the rule "*double x, and then add* $\frac{1}{2}$" would contain.

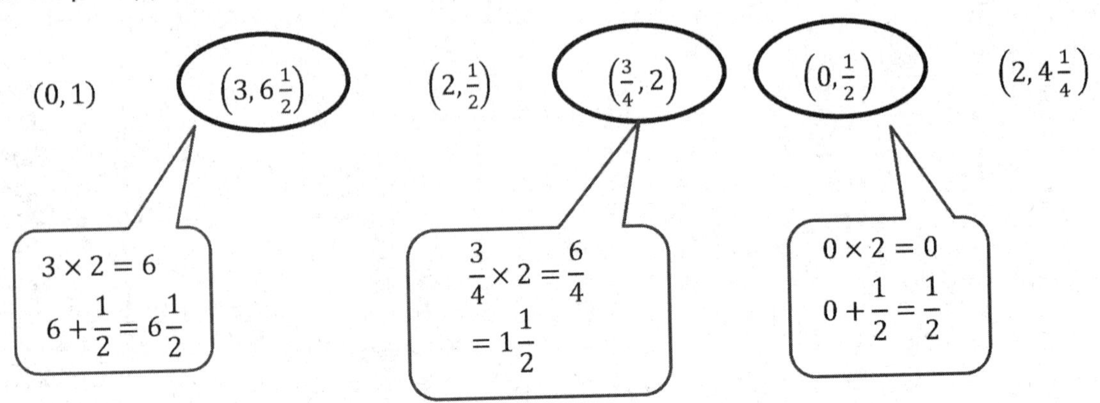

$3 \times 2 = 6$

$6 + \frac{1}{2} = 6\frac{1}{2}$

$\frac{3}{4} \times 2 = \frac{6}{4}$

$= 1\frac{1}{2}$

$0 \times 2 = 0$

$0 + \frac{1}{2} = \frac{1}{2}$

3. Give two other points that fall on this line.

$\left(\frac{1}{2}, 1\frac{1}{2}\right)$ $\left(1, 2\frac{1}{2}\right)$

I choose values for the x-coordinates. Then I doubled them and added $\frac{1}{2}$ to get the y-coordinates.

Name ______________________________ Date ______________

1. Complete the tables for the given rules.

Line ℓ

Rule: *Double x*

x	y	(x, y)
1		
2		
3		

Line m

Rule: *Double x, and then subtract 1*

x	y	(x, y)
1		
2		
3		

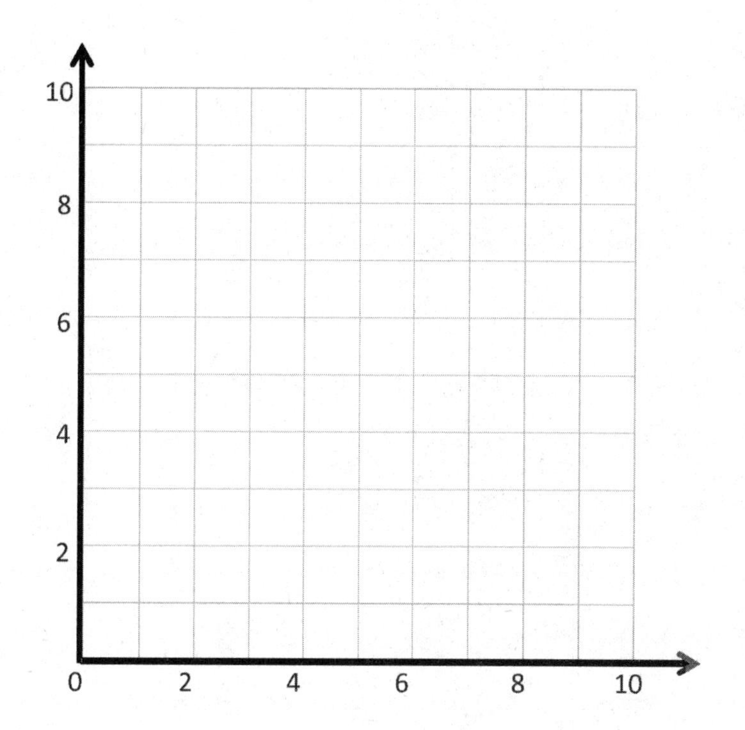

a. Draw each line on the coordinate plane above.

b. Compare and contrast these lines.

c. Based on the patterns you see, predict what the line for the rule *double x, and then add 1* would look like. Draw your prediction on the plane above.

2. Circle the point(s) that the line for the rule *multiply x by* $\frac{1}{2}$, *and then add 1* would contain.

$(0, \frac{1}{2})$ $(2, 1\frac{1}{4})$ $(2, 2)$ $(3, \frac{1}{2})$

a. Explain how you know.

b. Give two other points that fall on this line.

3. Complete the tables for the given rules.

Line ℓ

Rule: *Halve x, and then add 1*

x	y	(x, y)
0		
1		
2		
3		

Line m

Rule: *Halve x, and then add $1\frac{1}{4}$*

x	y	(x, y)
0		
1		
2		
3		

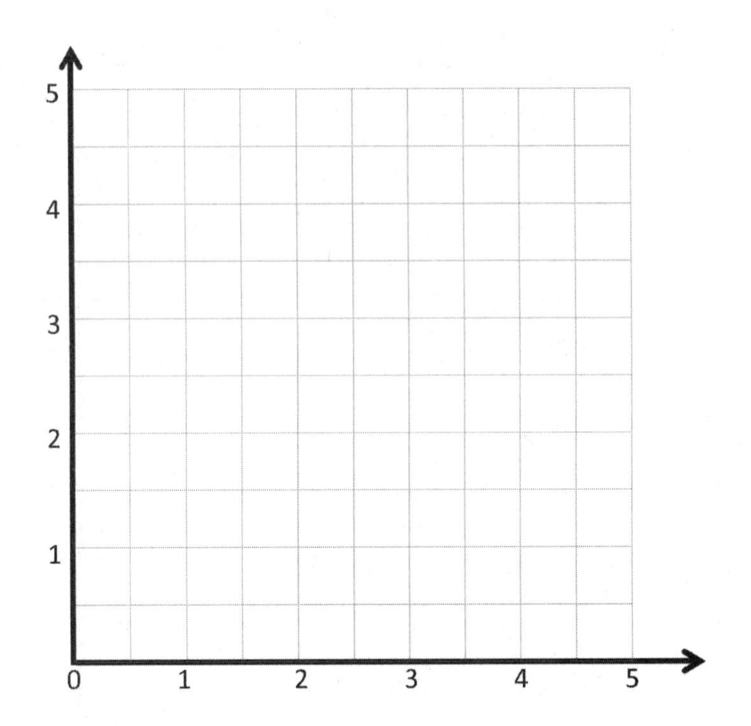

a. Draw each line on the coordinate plane above.

b. Compare and contrast these lines.

c. Based on the patterns you see, predict what the line for the rule *halve x, and then subtract 1* would look like. Draw your prediction on the plane above.

4. Circle the point(s) that the line for the rule *multiply x by $\frac{3}{4}$, and then subtract $\frac{1}{2}$* would contain.

$(1, \frac{1}{4})$ $(2, \frac{1}{4})$ $(3, 1\frac{3}{4})$ $(3, 1)$

a. Explain how you know.

b. Give two other points that fall on this line.

1. Write a rule for the line that contains the points $(0.3, 0.5)$ and $(1.0, 1.2)$.

   ***y* is 0.2 *more than* x.**

   a. Identify 2 more points on this line. Then draw it on the grid below.

Point	x	y	(x, y)
E	0.7	0.9	$(0.7, 0.9)$
F	1.5	1.7	$(1.5, 1.7)$

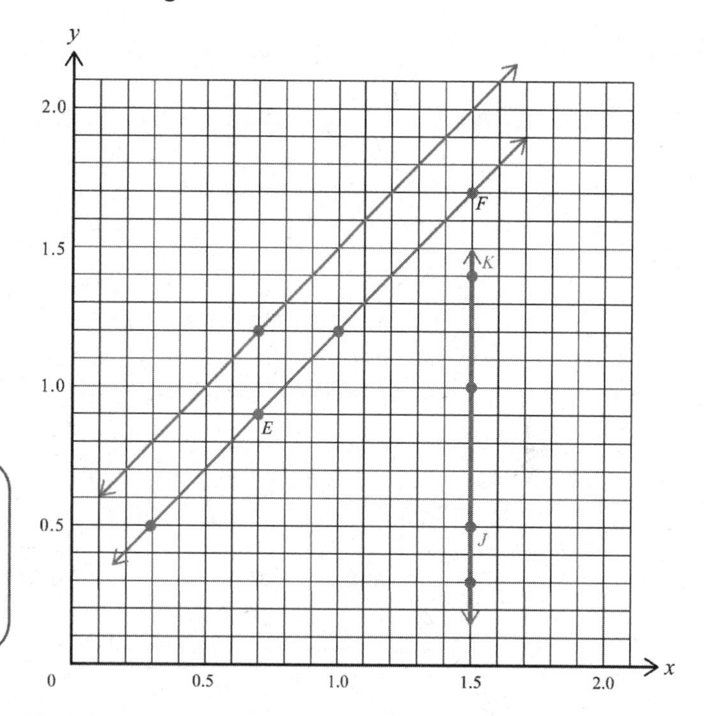

   b. Write a rule for a line that is parallel to $\overleftrightarrow{EF}$ and goes through point $(0.7, 1.2)$. Then draw the line on the grid.

   ***y* is 0.5 *more than* x.**

Since this line needs to be parallel to $\overleftrightarrow{EF}$, it must be an addition rule. In the coordinate pair $(0.7, 1.2)$, I can see that the y-coordinate is 0.5 more than the x-coordinate.

2. Give the rule for the line that contains the points $(1.5, 0.3)$ and $(1.5, 1.0)$.

   ***x* is always 1.5.**

   a. Identify 2 more points on this line. Draw the line on the grid above.

Point	x	y	(x, y)
J	1.5	0.5	$(1.5, 0.5)$
K	1.5	1.4	$(1.5, 1.4)$

   b. Write a rule for a line that is parallel to $\overleftrightarrow{JK}$.

   ***x* is always 1.8.**

Since this line must be parallel to $\overleftrightarrow{JK}$, it must be another vertical line where the x-coordinate is always the same.

3. Give the rule for a line that contains the point $(0.3, 0.9)$ using the operation or description below. Then, name 2 other points that would fall on each line.

a. Addition: ***y is 0.6 more than x.***

Point	x	y	(x,y)
T	0.4	1	$(0.4, 1)$
U	1	1.6	$(1, 1.6)$

b. A line parallel to the x-axis: ***y is always 0.9.***

Point	x	y	(x,y)
G	0.4	0.9	$(0.4, 0.9)$
H	1	0.9	$(1, 0.9)$

A line parallel to the x-axis is a horizontal line. Horizontal lines have y-coordinates that do not change.

c. Multiplication: ***y is x tripled.***

Point	x	y	(x,y)
A	0.2	0.6	$(0.2, 0.6)$
B	0.5	1.5	$(0.5, 1.5)$

d. A line parallel to the y-axis: ***x is always 0.3.***

Point	x	y	(x,y)
V	0.3	1.3	$(0.3, 1.3)$
W	0.3	2	$(0.3, 2)$

A line parallel to the y-axis is a vertical line. Vertical lines have x-coordinates that do not change.

e. Multiplication with addition: ***Double x, and then add 0.3.***

Point	x	y	(x,y)
R	0.4	1.1	$(0.4, 1.1)$
S	0.5	1.3	$(0.5, 1.3)$

I can use the original coordinate pair, $(0.3, 0.9)$, to help me generate a multiplication with addition rule.

$0.3 \times 2 = 0.6$ (This is the "*Double* x" part of the rule.)

$0.6 + 0.3 = 0.9$ (This is the "*then add* 0.3" part of the rule.)

Name ______________________________ Date ______________

1. Write a rule for the line that contains the points $(0, \frac{1}{4})$ and $(2\frac{1}{2}, 2\frac{3}{4})$.

   a. Identify 2 more points on this line. Draw the line on the grid below.

Point	x	y	(x, y)
B			
C			

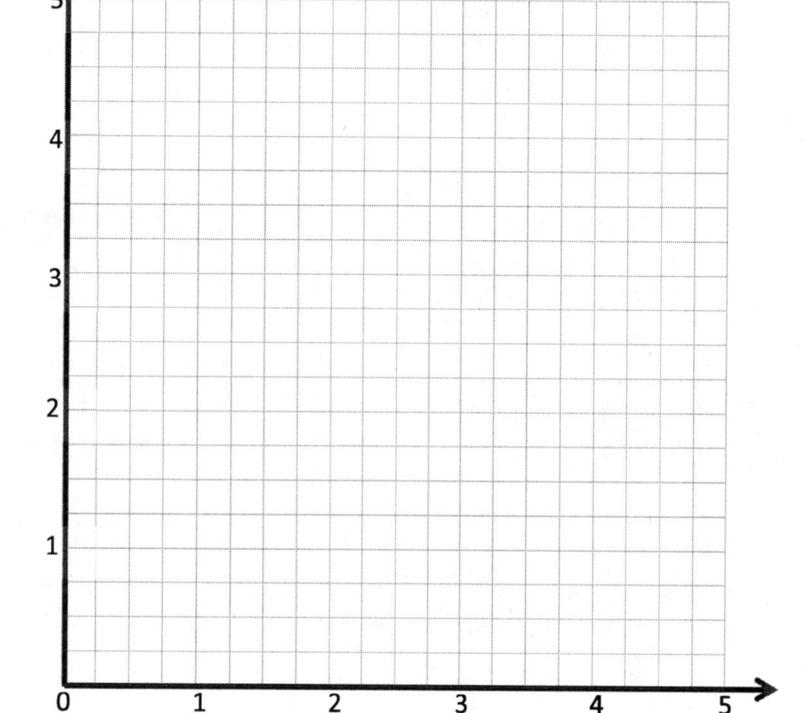

   b. Write a rule for a line that is parallel to $\overleftrightarrow{BC}$ and goes through point $(1, 2\frac{1}{4})$.

2. Give the rule for the line that contains the points $(1, 2\frac{1}{2})$ and $(2\frac{1}{2}, 2\frac{1}{2})$.

   a. Identify 2 more points on this line. Draw the line on the grid above.

Point	x	y	(x, y)
G			
H			

   b. Write a rule for a line that is parallel to $\overleftrightarrow{GH}$.

3. Give the rule for a line that contains the point $(\frac{3}{4}, 1\frac{1}{2})$ using the operation or description below. Then, name 2 other points that would fall on each line.

a. Addition: ______________

Point	x	y	(x, y)
T			
U			

b. A line parallel to the x-axis: ______________

Point	x	y	(x, y)
G			
H			

c. Multiplication: ______________

Point	x	y	(x, y)
A			
B			

d. A line parallel to the y-axis: ______________

Point	x	y	(x, y)
V			
W			

e. Multiplication with addition: ______________

Point	x	y	(x, y)
R			
S			

4. On the grid, two lines intersect at (1.2, 1.2). If line a passes through the origin and line b contains the point (1.2, 0), write a rule for line a and line b.

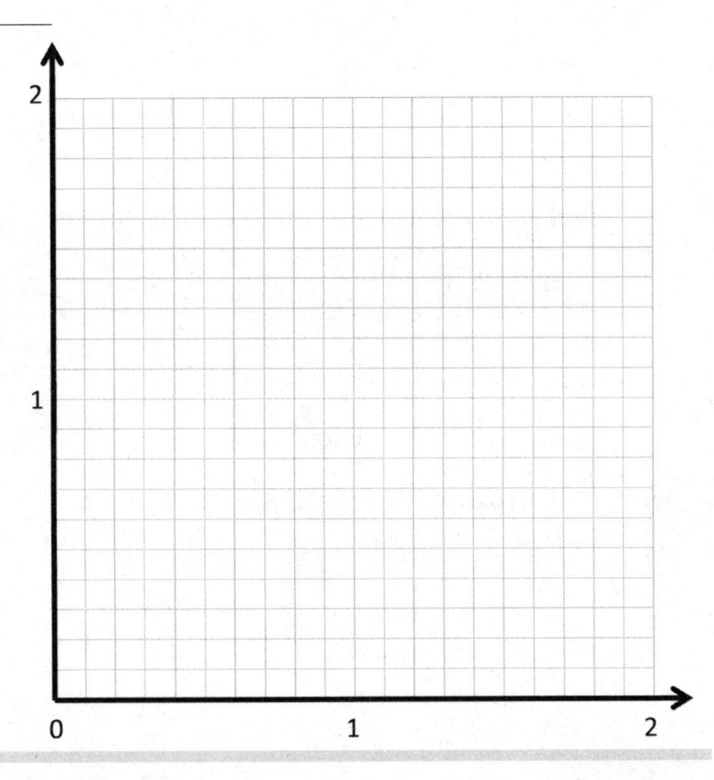

1. Maya and Ruvio used their right angle templates and straightedges to draw sets of parallel lines. Who drew a correct set of parallel lines and why?

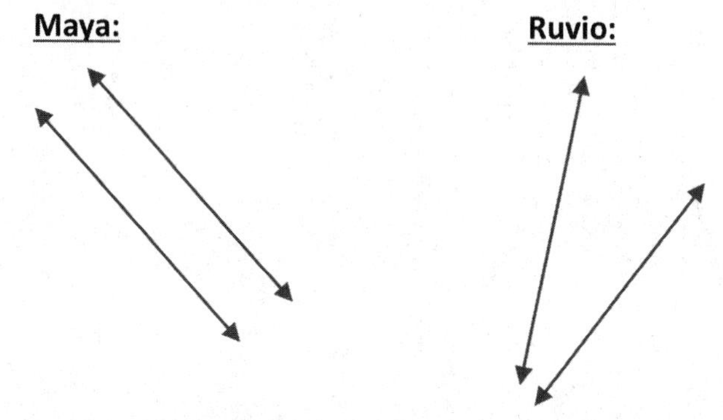

Maya drew a correct set of parallel lines because if you extend her lines, they will never intersect (cross). If you extend Ruvio's lines, they will intersect.

2. On the grid below, Maya circled all the sets of segments that she thought were parallel. Is she correct? Why or why not?

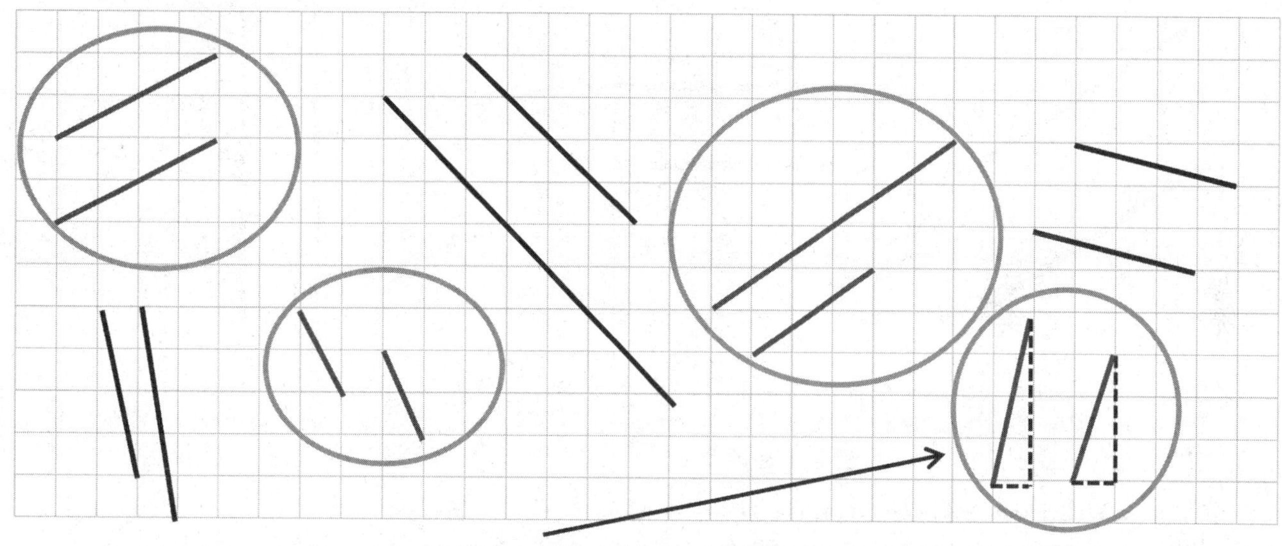

Maya is not completely correct. This set is not parallel. I drew a horizontal and vertical dotted line near each segment to complete a triangle. Even though both triangles have a base of 1, the left triangle is taller. I can see that if I were to extend these segments, they would eventually intersect. These segments are not parallel. Also, Maya did not circle all of the parallel sets of segments.

3. Use your straightedge to draw a segment parallel to each segment through the given point.

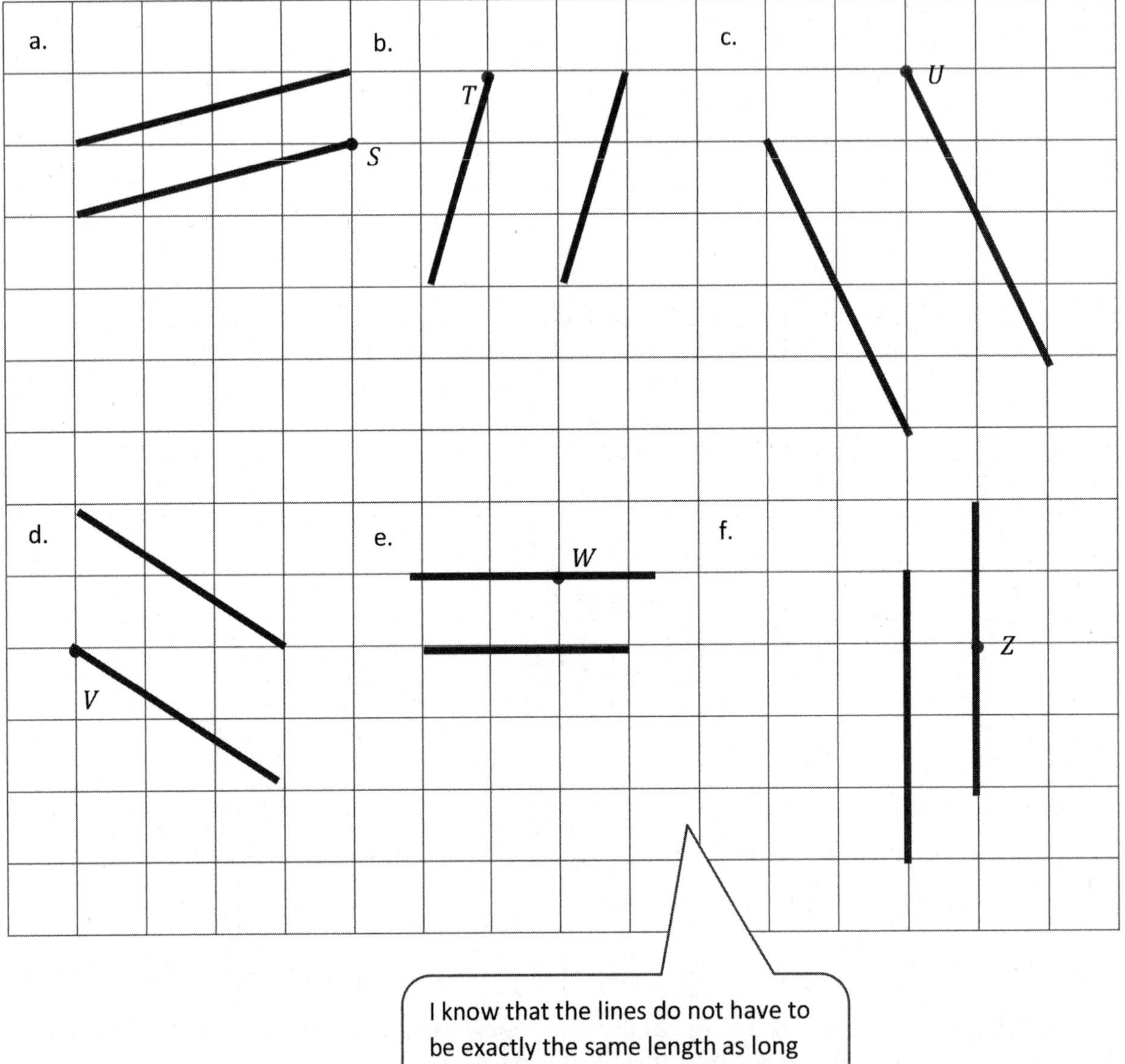

Name ______________________________ Date ______________

1. Use your right angle template and straightedge to draw at least three sets of parallel lines in the space below.

2. Circle the segments that are parallel.

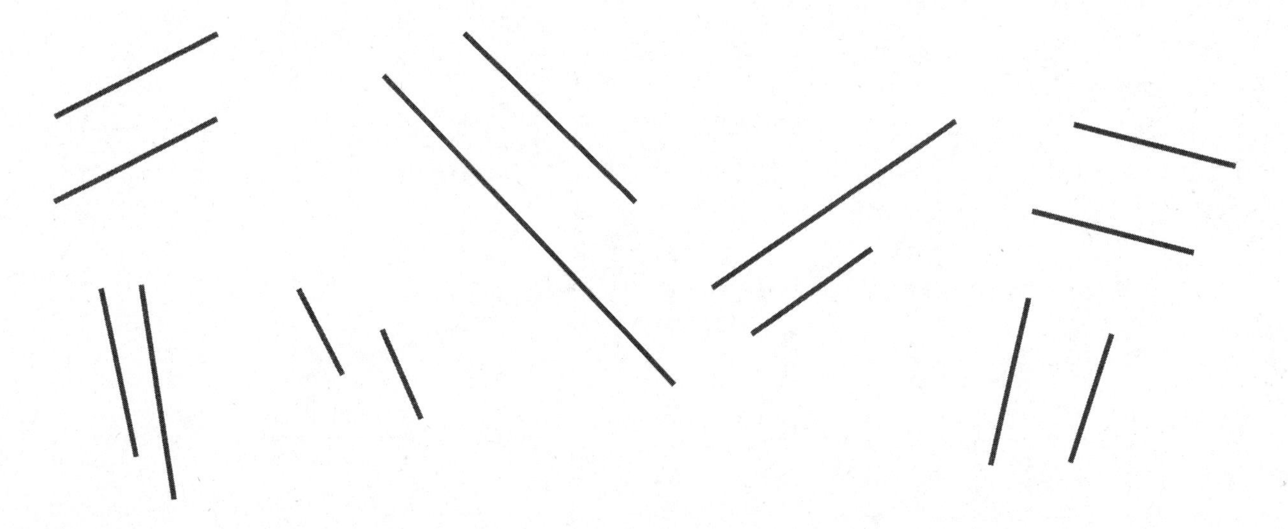

3. Use your straightedge to draw a segment parallel to each segment through the given point.

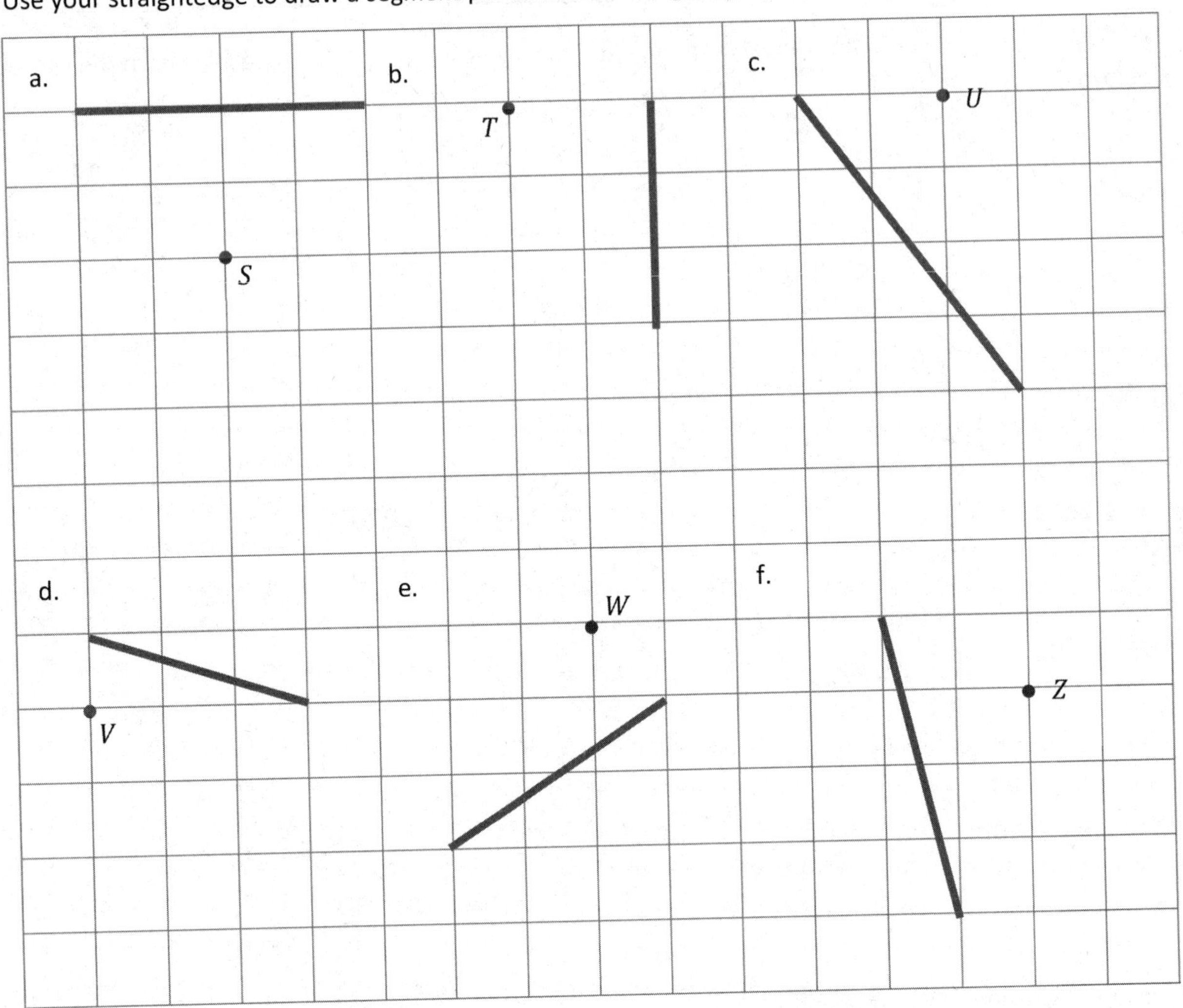

4. Draw 2 different lines parallel to line *b*.

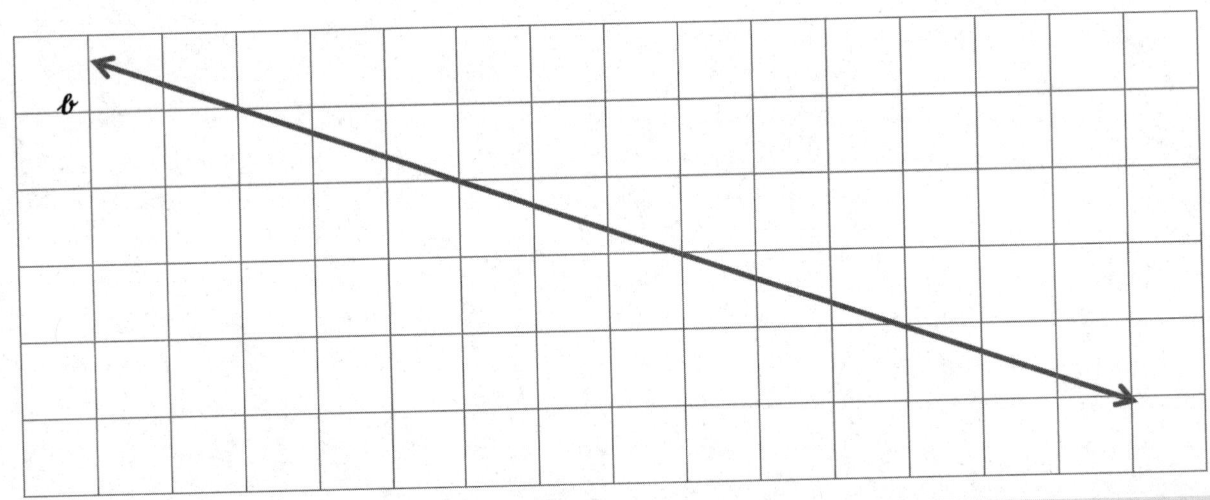

1. Use the coordinate plane below to complete the following tasks.

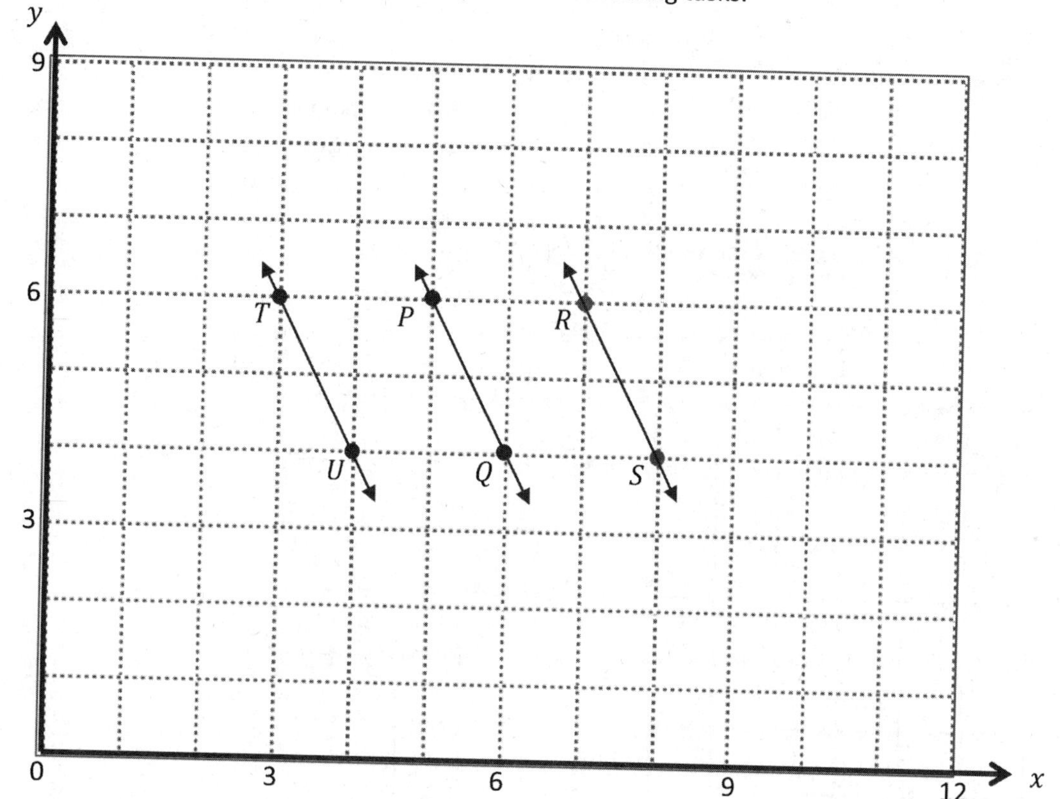

a. Identify the locations of P and Q. P (5 , 6) Q (6 , 4)

b. Draw $\overleftrightarrow{PQ}$.

The symbol ⊥ means perpendicular.
The symbol ‖ means parallel.

c. Plot the following coordinate pairs on the plane: R $(7, 6)$ S $(8, 4)$

d. Draw $\overleftrightarrow{RS}$.

e. Circle the relationship between $\overleftrightarrow{PQ}$ and $\overleftrightarrow{RS}$. $\overleftrightarrow{PQ} \perp \overleftrightarrow{RS}$

f. Give the coordinates of a pair of points, T and U, such that $\overleftrightarrow{TU} \parallel \overleftrightarrow{PQ}$.

T (**3** , **6**) U (**4** , **4**)

There are many possible sets of coordinates that would make $\overleftrightarrow{TU}$ parallel to $\overleftrightarrow{PQ}$. I can keep the y-coordinates the same and move the x-coordinates 2 units to the left.

g. Draw $\overleftrightarrow{TU}$.

2. Use the coordinate plane below to complete the following tasks.

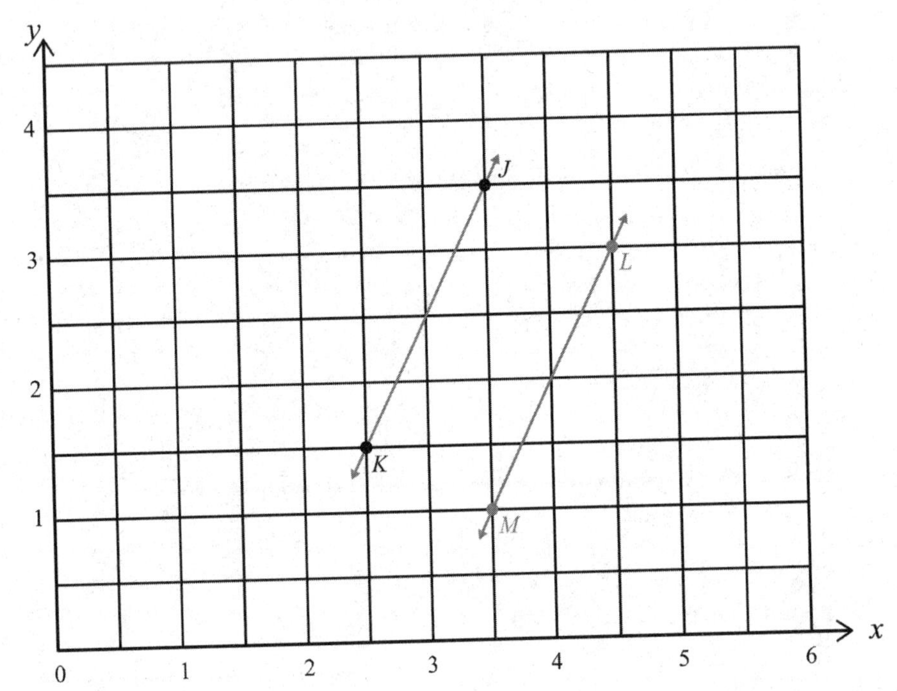

a. Identify the locations of J and K. J $\left(3\frac{1}{2}, 3\frac{1}{2}\right)$ K $\left(2\frac{1}{2}, 1\frac{1}{2}\right)$

b. Draw $\overleftrightarrow{JK}$.

c. Generate coordinate pairs for L and M such tha $\overleftrightarrow{JK} \parallel \overleftrightarrow{LM}$. L $\left(4\frac{1}{2}, 3\right)$ M $\left(3\frac{1}{2}, 1\right)$

d. Draw $\overleftrightarrow{LM}$.

e. Explain the pattern you used when generating coordinate pairs for L and M.

I visualized shifting points J and K one unit to the <u>right</u>, which is two grid lines. As a result, the x-coordinates of L and M are 1 greater than those of J and K.

Then I visualized shifting both points <u>down</u> one-half unit, which is one grid line. As a result, the y-coordinates of L and M are $\frac{1}{2}$ less than those of J and K.

Name ______________________________ Date ______________

1. Use the coordinate plane below to complete the following tasks.

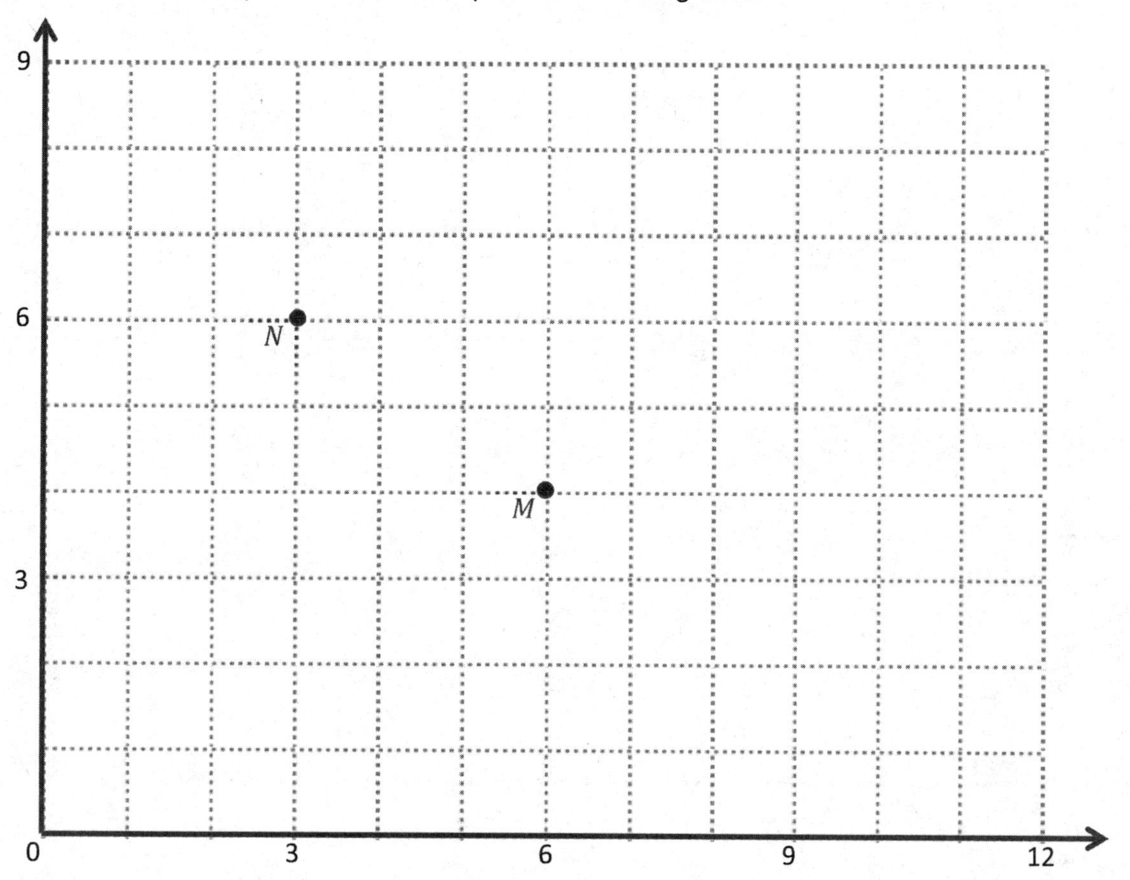

a. Identify the locations of M and N. M: (____, ____) N: (____, ____)

b. Draw $\overleftrightarrow{MN}$.

c. Plot the following coordinate pairs on the plane.

J: (5, 7) K: (8, 5)

d. Draw $\overleftrightarrow{JK}$.

e. Circle the relationship between $\overleftrightarrow{MN}$ and $\overleftrightarrow{JK}$. $\overleftrightarrow{MN} \perp \overleftrightarrow{JK}$ $\overleftrightarrow{MN} \parallel \overleftrightarrow{JK}$

f. Give the coordinates of a pair of points, F and G, such that $\overleftrightarrow{FG} \parallel \overleftrightarrow{MN}$.

F: (____, ____) G: (____, ____)

g. Draw $\overleftrightarrow{FG}$.

2. Use the coordinate plane below to complete the following tasks.

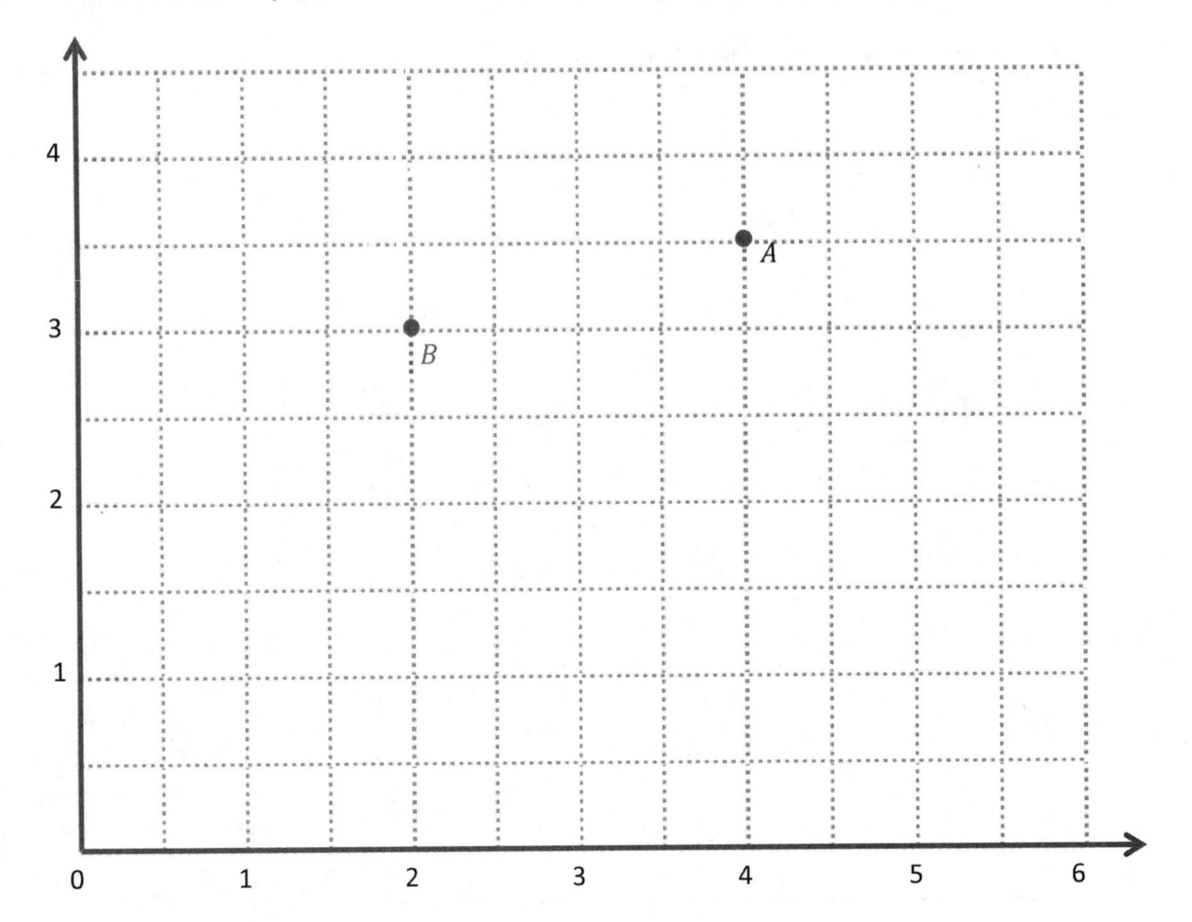

a. Identify the locations of A and B. A: (____, ____) B: (____, ____)

b. Draw $\overleftrightarrow{AB}$.

c. Generate coordinate pairs for C and D, such that $\overleftrightarrow{AB} \parallel \overleftrightarrow{CD}$.

C: (____, ____) D: (____, ____)

d. Draw $\overleftrightarrow{CD}$.

e. Explain the pattern you used when generating coordinate pairs for C and D.

f. Give the coordinates of a point, F, such that $\overleftrightarrow{AB} \parallel \overleftrightarrow{EF}$.

E: $(2\frac{1}{2}, 2\frac{1}{2})$ F: (____, ____)

g. Explain how you chose the coordinates for F.

Perpendicular segments intersect and form 90°, or right, angles.

1. Circle the pairs of segments that are perpendicular.

The angle formed by these segments is greater than 90°. These segments are *not* perpendicular.

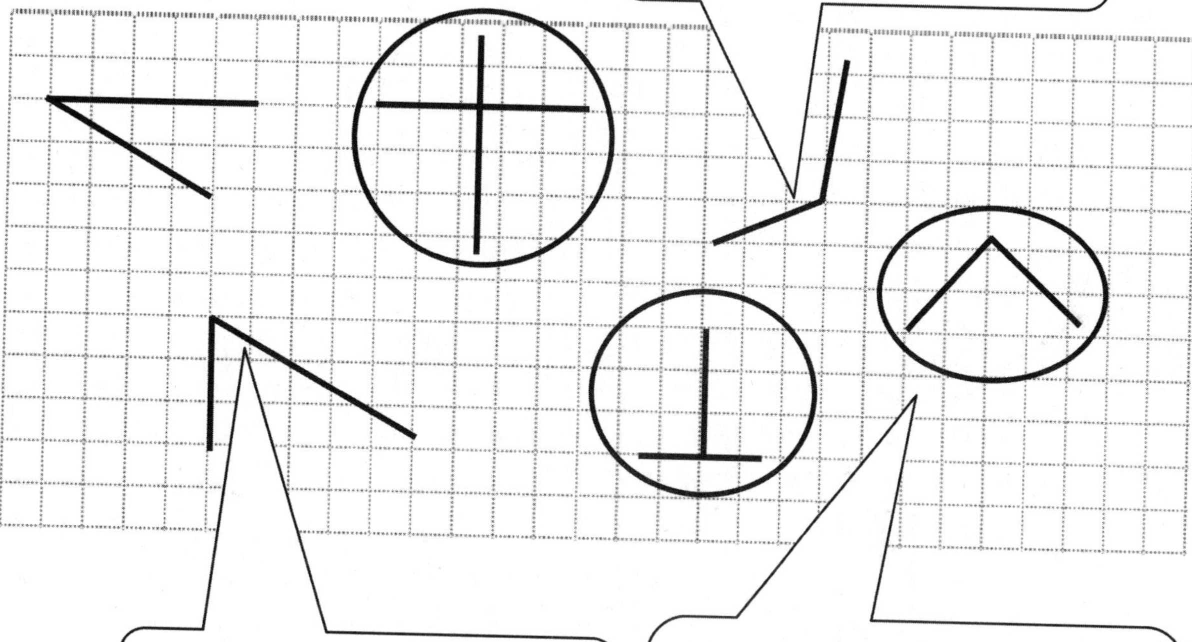

The angle formed by these segments is less than 90°. These segments are *not* perpendicular.

I can use anything that is a right angle, such as the corner of a paper, to see if it fits in the angle where the lines intersect. If it fits perfectly, then I know that the lines are perpendicular.

2. Draw a segment perpendicular to each given segment. Show your thinking by sketching triangles as needed.

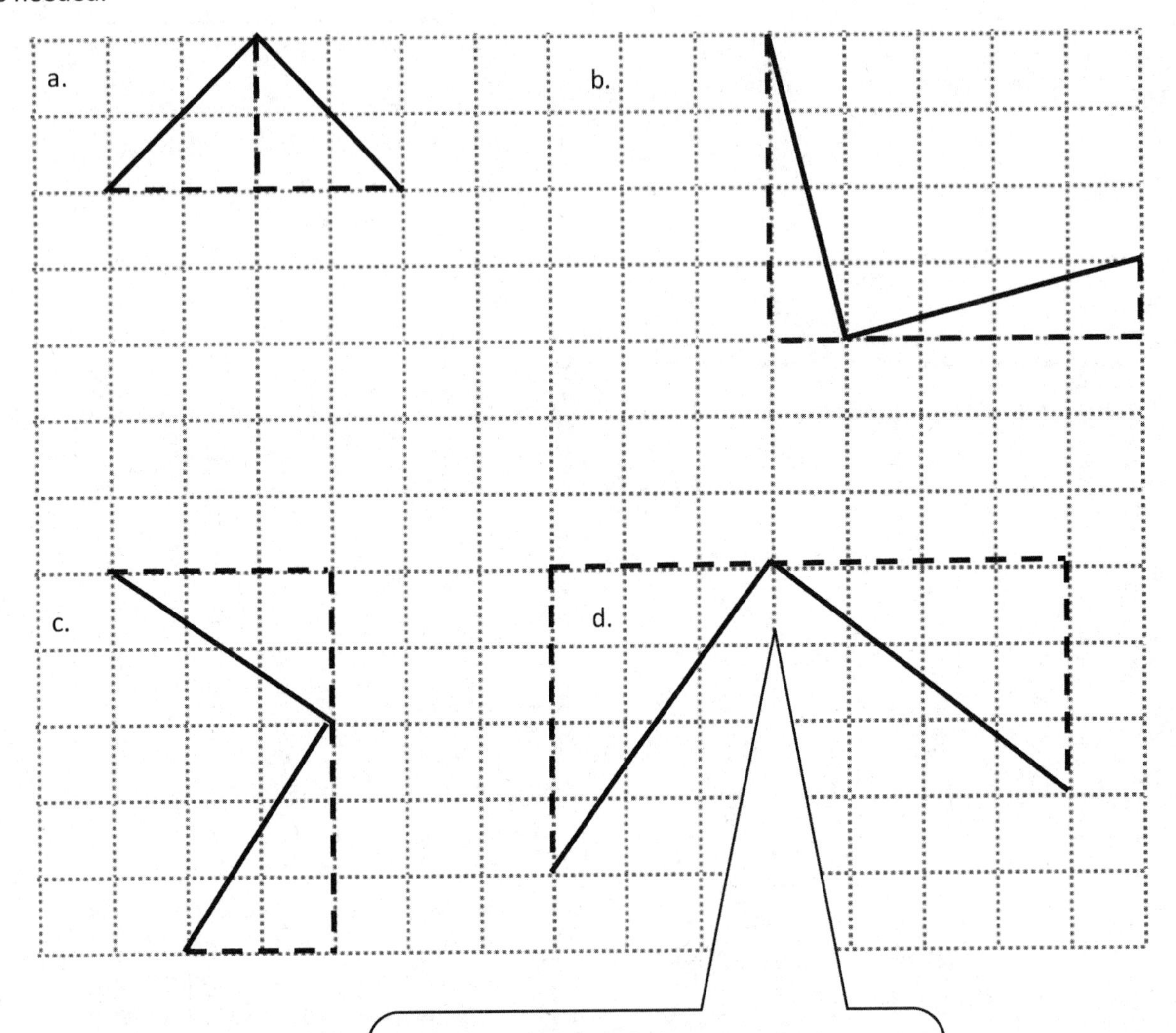

Name ______________________________ Date ______________

1. Circle the pairs of segments that are perpendicular.

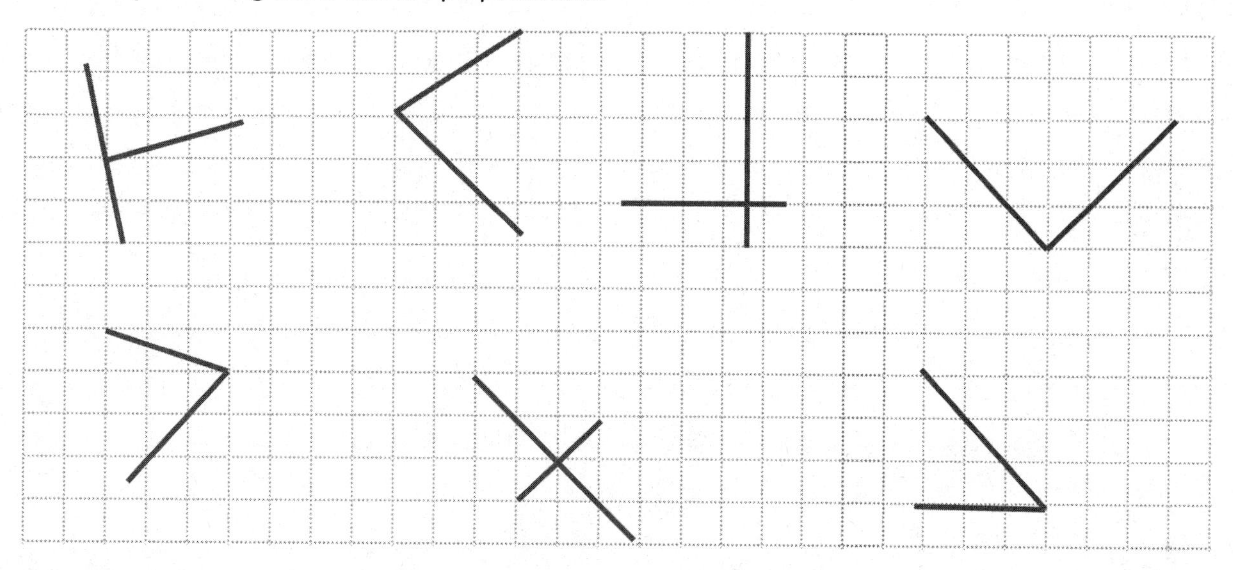

2. In the space below, use your right triangle templates to draw at least 3 different sets of perpendicular lines.

3. Draw a segment perpendicular to each given segment. Show your thinking by sketching triangles as needed.

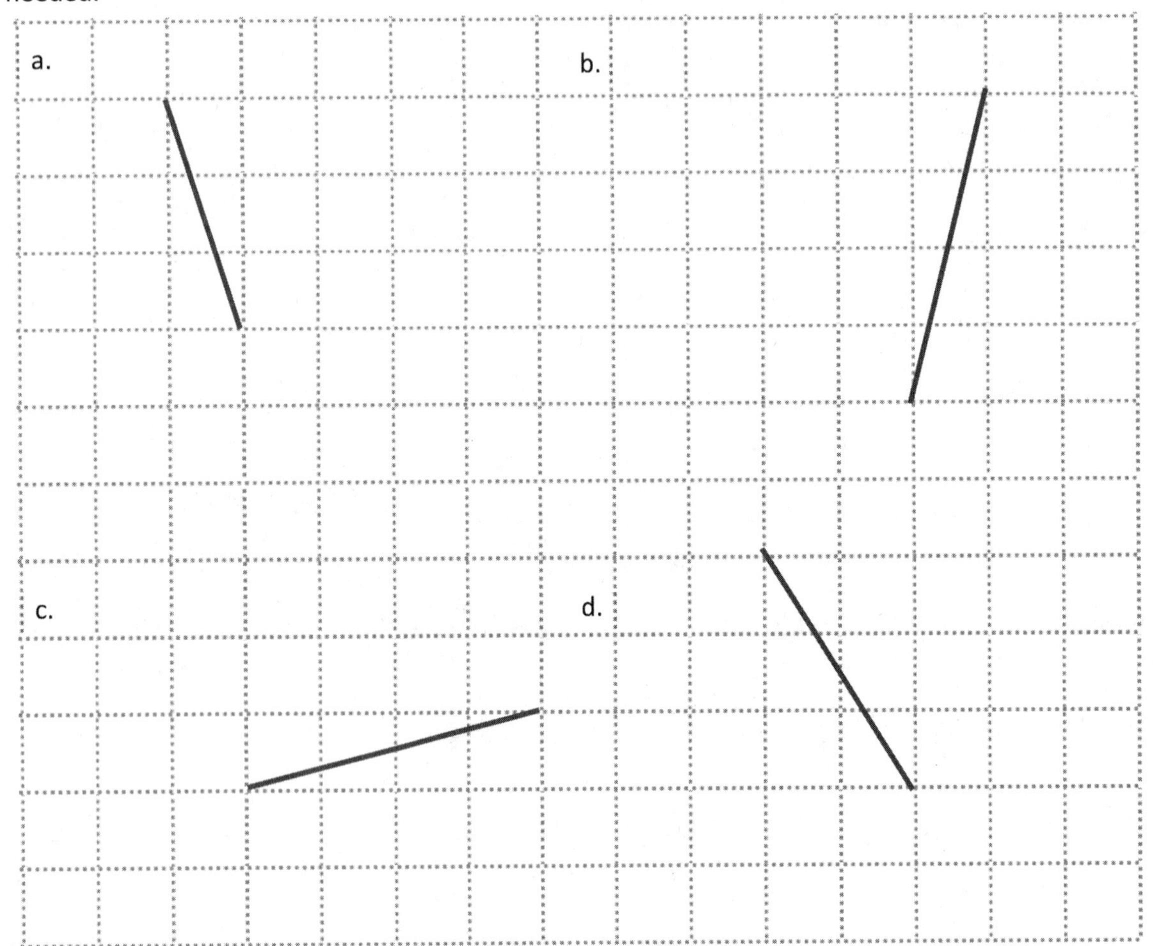

4. Draw 2 different lines perpendicular to line b.

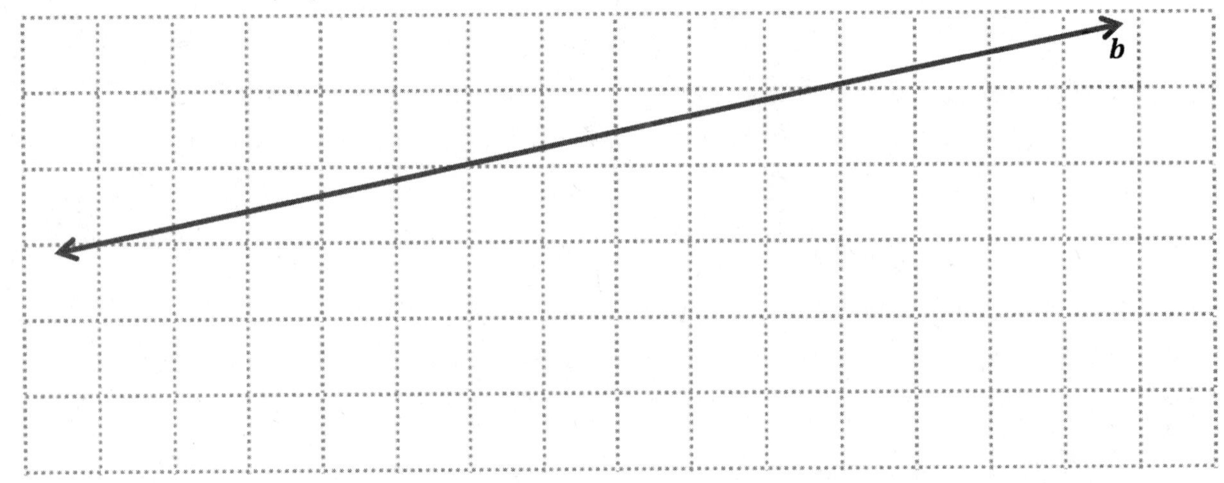

1. In the right triangle below, the measure of angle L is 50°. What is the measure of angle K?

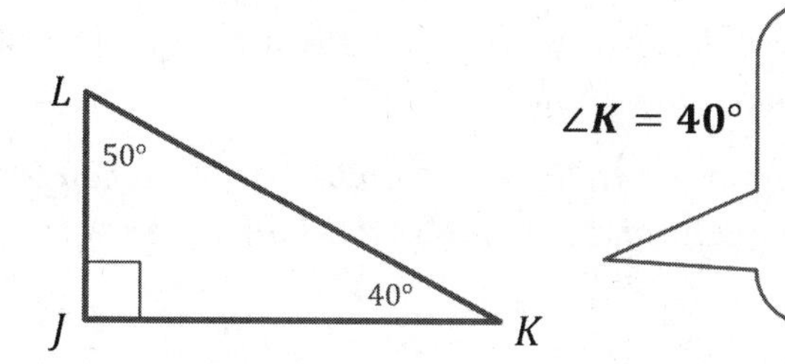

$\angle K = 40°$

The sum of the interior angles of *all* triangles is 180°. Triangle JKL is a right triangle. Since $\angle J$ is 90°, and $\angle L$ is 50°, $\angle K$ must be 40°.

$180° - 90° - 50° = 40°$

2. Use the coordinate plane below to complete the following tasks.

   a. Draw $\overline{KL}$.

   b. Plot point $(5, 8)$.

   c. Draw $\overline{LM}$.

After I sketch the right triangle, I can visualize it sliding and rotating. These triangles are the same.

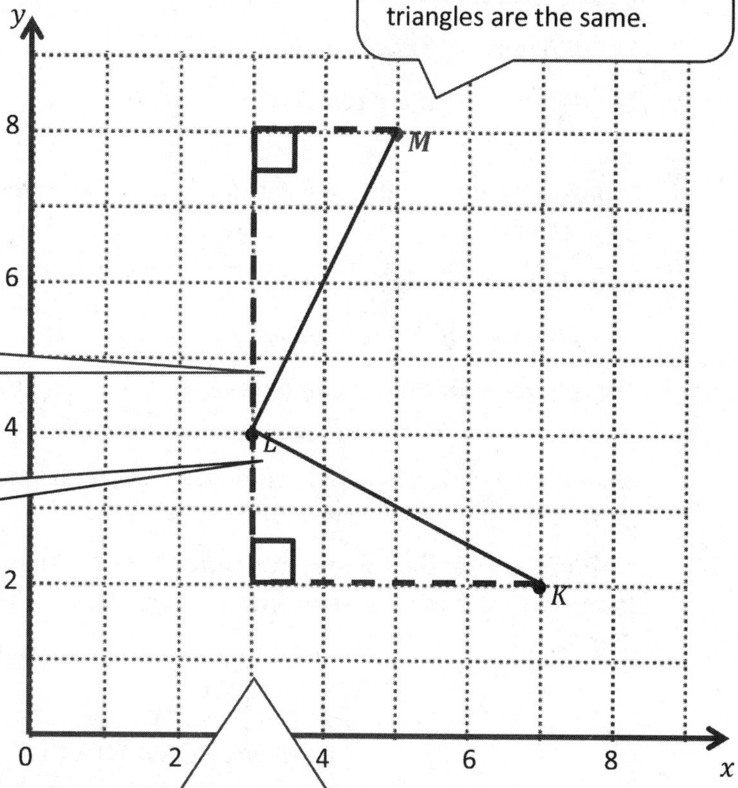

This is an acute angle, like $\angle K$, in Problem 1.

This is an acute angle, like $\angle L$, in Problem 1.

The two triangles I sketched are aligned to create a 180°, or straight angle, along the vertical grid line. So if the two acute angles of the triangles add up to 90°, the angle in between them, $\angle MLK$, must also be 90°.

d. Explain how you know $\angle MLK$ is a right angle without measuring it.

I used the grid lines to sketch a right triangle with side $\overline{LK}$, just like in Problem 1. Then I visualized sliding and rotating the triangle so side $\overline{LK}$ matched up with side $\overline{LM}$.

I know that the measures of the 2 acute angles of a right triangle add up to $90°$. So when the long side of the triangle and the short side of the triangle form a straight angle, $180°$, the angle in between them, $\angle MLK$, is also $90°$.

e. Compare the coordinates of points L and K. What is the difference of the x-coordinates? The y-coordinates?

$L\ (3, 4)$ and $K\ (7, 2)$

The difference of the x-coordinates is 4.

The difference of the y-coordinates is 2.

f. Compare the coordinates of points L and M. What is the difference of the x-coordinates? The y-coordinates?

$L\ (3, 4)$ and $M\ (5, 8)$

The difference of the x-coordinates is 2.

The difference of the y-coordinates is 4.

g. What is the relationship of the differences you found in parts (e) and (f) to the triangles of which these two segments are a part?

The difference in the value of the coordinates is either 2 or 4. That makes sense to me because the triangles that these two segments are part of have a height of either 2 or 4 and a base of either 2 or 4.

When I visualize the triangle sliding and rotating, it makes sense that the x-coordinates and y-coordinates will change by a value of 2 or 4 because that's the length of the triangle's height and base.

Name ____________________ Date ____________

1. Use the coordinate plane below to complete the following tasks.

   a. Draw $\overline{PQ}$.

   b. Plot point R (3, 8).

   c. Draw $\overline{PR}$.

   d. Explain how you know $\angle RPQ$ is a right angle without measuring it.

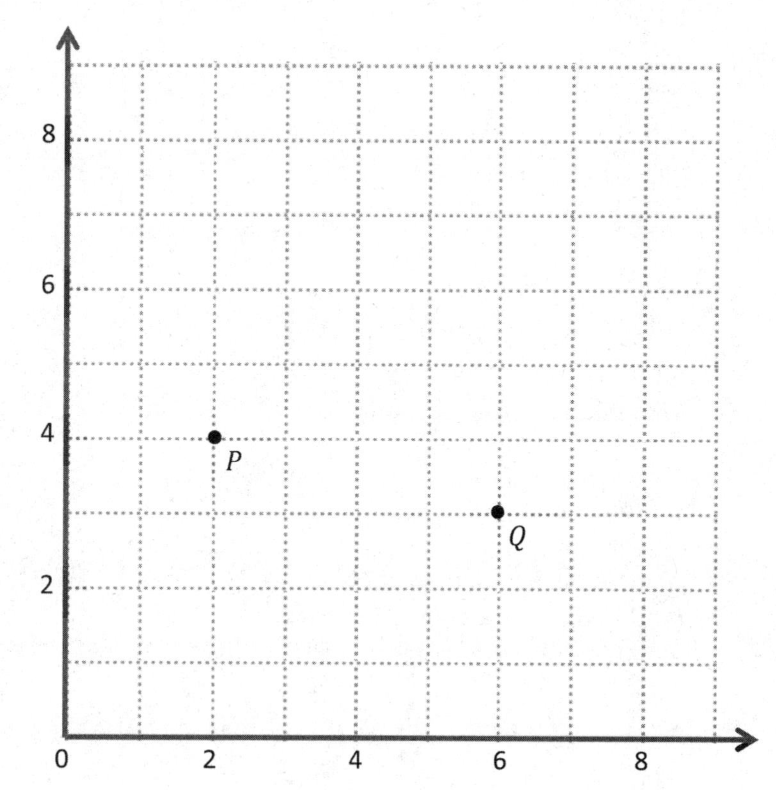

   e. Compare the coordinates of points P and Q. What is the difference of the x-coordinates? The y-coordinates?

   f. Compare the coordinates of points P and R. What is the difference of the x-coordinates? The y-coordinates?

   g. What is the relationship of the differences you found in parts (e) and (f) to the triangles of which these two segments are a part?

2. Use the coordinate plane below to complete the following tasks.

   a. Draw $\overline{CB}$.

   b. Plot point $D\left(\frac{1}{2}, 5\frac{1}{2}\right)$.

   c. Draw $\overline{CD}$.

   d. Explain how you know $\angle DCB$ is a right angle without measuring it.

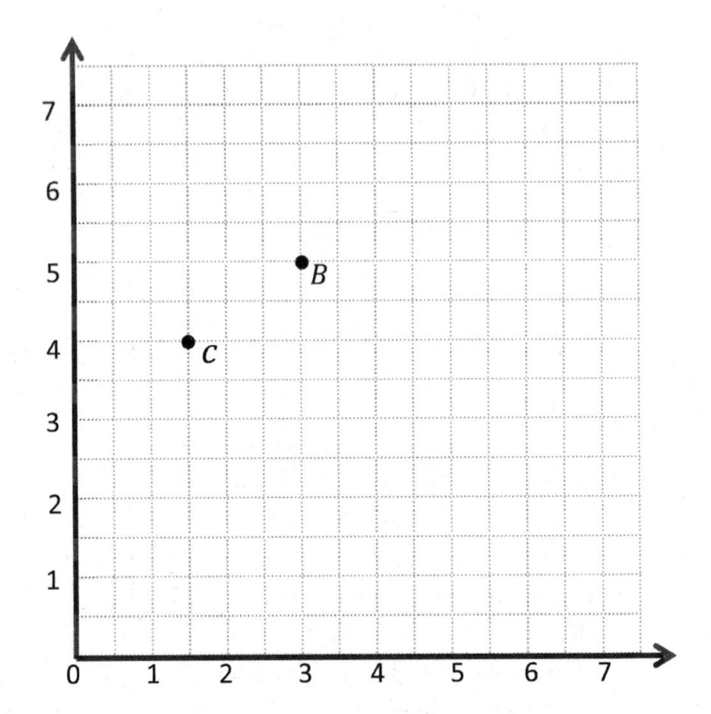

   e. Compare the coordinates of points C and B. What is the difference of the x-coordinates? The y-coordinates?

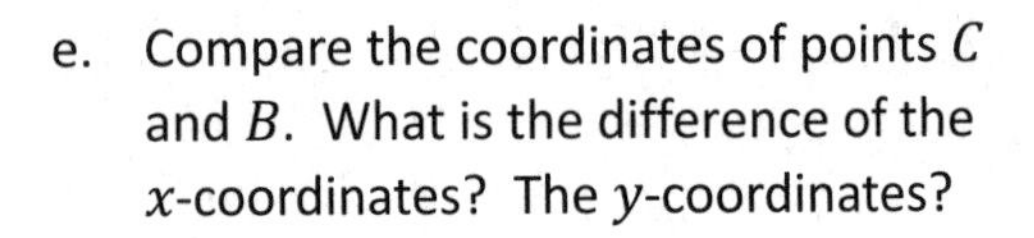

   f. Compare the coordinates of points C and D. What is the difference of the x-coordinates? The y-coordinates?

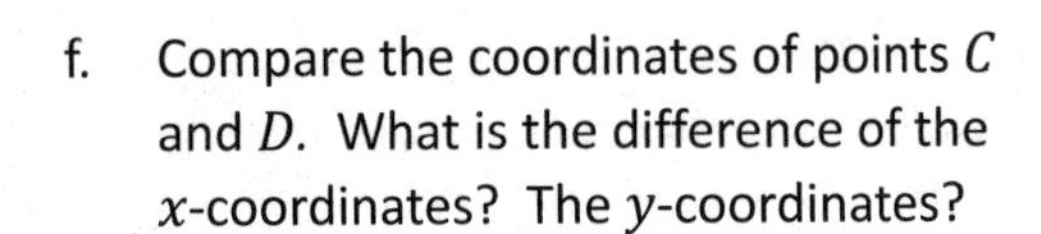

   g. What is the relationship of the differences you found in parts (e) and (f) to the triangles of which these two segments are a part?

3. $\overleftrightarrow{ST}$ contains the following points. S: (2, 3) T: (9, 6)

   Give the coordinates of a pair of points, U and V, such that $\overleftrightarrow{ST} \perp \overleftrightarrow{UV}$.

   U: (____, ____) V: (____, ____)

In order to create a figure that is symmetric about $\overleftrightarrow{UR}$, I need to find points that are drawn using a line *perpendicular to* and *equidistant from* (the same distance from) the line of symmetry, $\overleftrightarrow{UR}$.

1. Draw to create a figure that is symmetric about $\overleftrightarrow{UR}$.

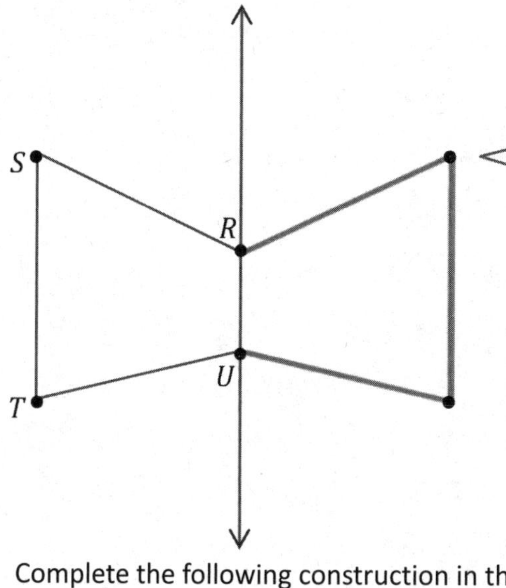

The distance from this point to the line of symmetry is the same as the distance from the line of symmetry to point S, when measured on a line perpendicular to the line of symmetry.

I know that collinear means that the points are "lying on the same straight line," so non-collinear must mean that the three points are *not* on the same straight line.

2. Complete the following construction in the space below.
   a. Plot 3 non-collinear points, A, B, and C.
   b. Draw $\overline{AB}$, $\overline{BC}$, and $\overleftrightarrow{AC}$.
   c. Plot point D, and draw the remaining sides, such that quadrilateral $ABCD$ is symmetric about $\overleftrightarrow{AC}$.

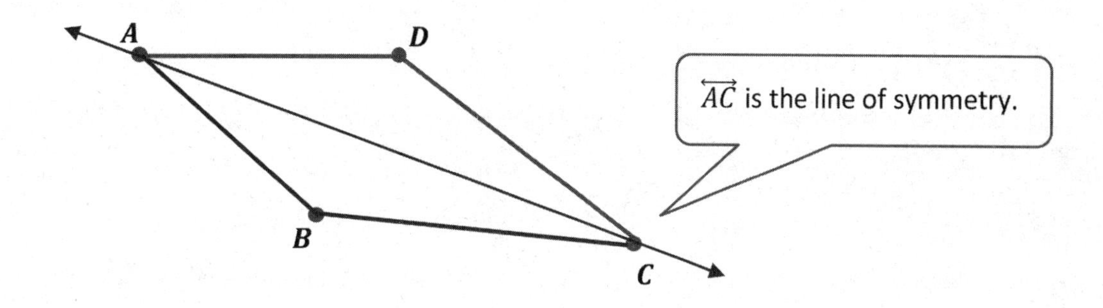

$\overleftrightarrow{AC}$ is the line of symmetry.

Name ______________________________ Date ______________

1. Draw to create a figure that is symmetric about $\overleftrightarrow{DE}$.

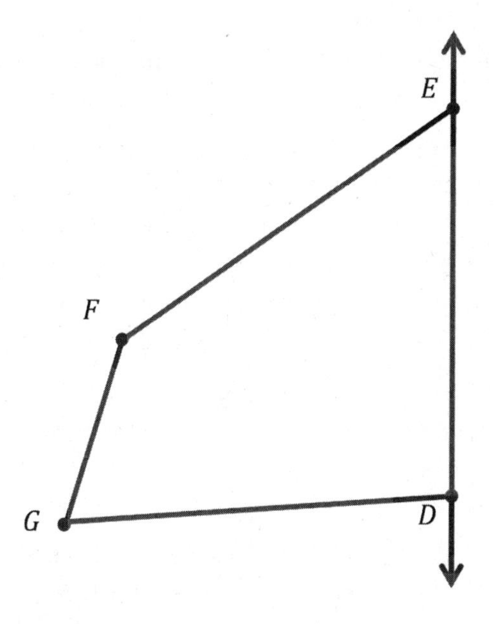

2. Draw to create a figure that is symmetric about $\overleftrightarrow{LM}$.

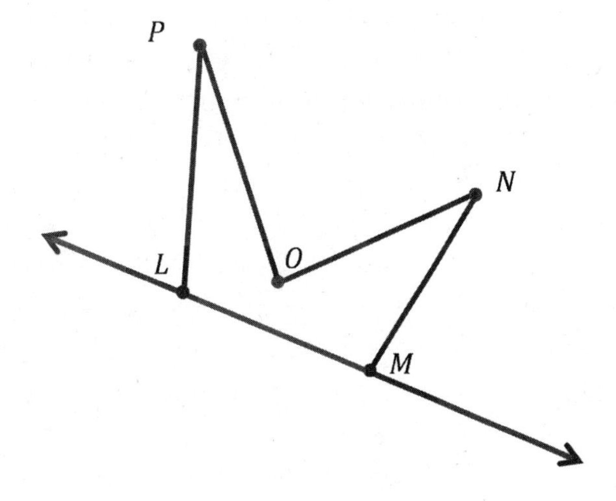

3. Complete the following construction in the space below.

   a. Plot 3 non-collinear points, G, H, and I.

   b. Draw $\overline{GH}$, $\overline{HI}$, and $\overleftrightarrow{IG}$.

   c. Plot point J, and draw the remaining sides, such that quadrilateral $GHIJ$ is symmetric about $\overleftrightarrow{IG}$.

4. In the space below, use your tools to draw a symmetric figure about a line.

Use the plane to the right to complete the following tasks.

This will be a vertical line.

a. Draw a line h whose rule is x is always 7.

b. Plot the points from Table A on the grid in order. Then, draw line segments to connect the points in order.

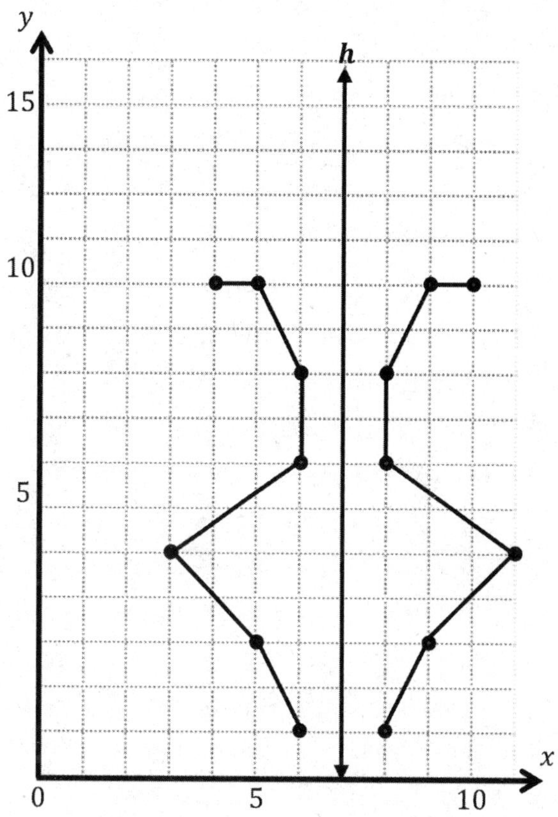

Table A

(x, y)
$(6, 1)$
$(5, 3)$
$(3, 5)$
$(6, 7)$
$(6, 7)$
$(5, 11)$
$(4, 11)$

Table B

(x, y)
$(8, 1)$
$(9, 3)$
$(11, 5)$
$(8, 7)$
$(8, 9)$
$(9, 11)$
$(10, 11)$

c. Complete the drawing to create a figure that is symmetric about line h. For each point in Table A, record the symmetric point on the other side of h.

d. Compare the y-coordinates in Table A with those in Table B. What do you notice?

The y-coordinates in Table A are the same as in Table B. Because the line of symmetry is a vertical ine, only the x-coordinates will change.

e. Compare the x-coordinates in Table A with those in Table B. What do you notice?

I notice that the difference in the x-coordinates is always an even number because the distance that a point is from line h has to double.

Name ______________________________ Date ______________

1. Use the plane to the right to complete the following tasks.

   a. Draw a line s whose rule is x *is always 5*.

   b. Plot the points from Table A on the grid in order. Then, draw line segments to connect the points in order.

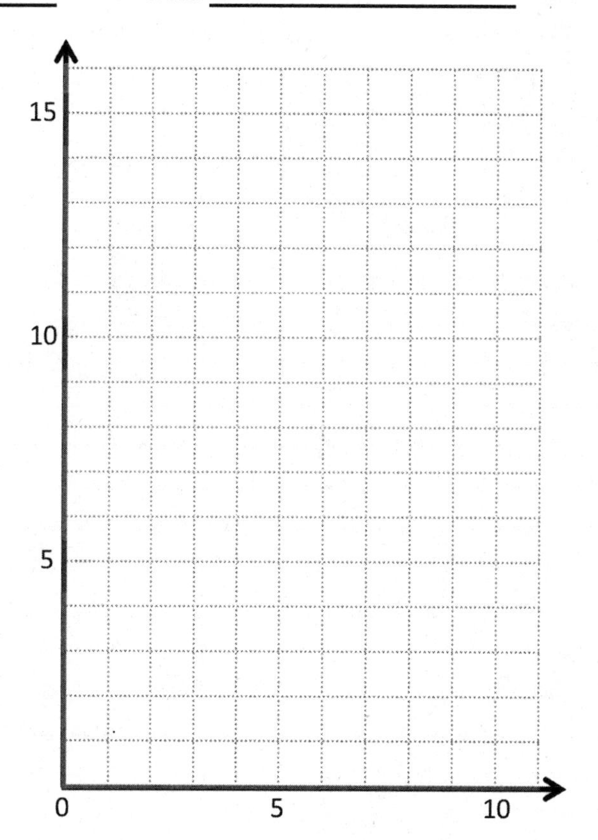

Table A

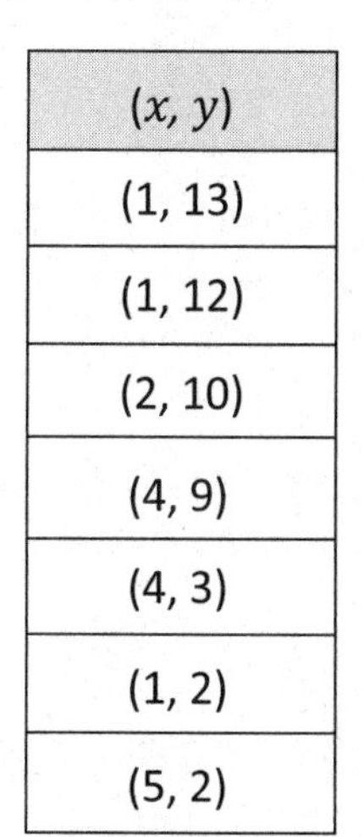

(x, y)
(1, 13)
(1, 12)
(2, 10)
(4, 9)
(4, 3)
(1, 2)
(5, 2)

Table B

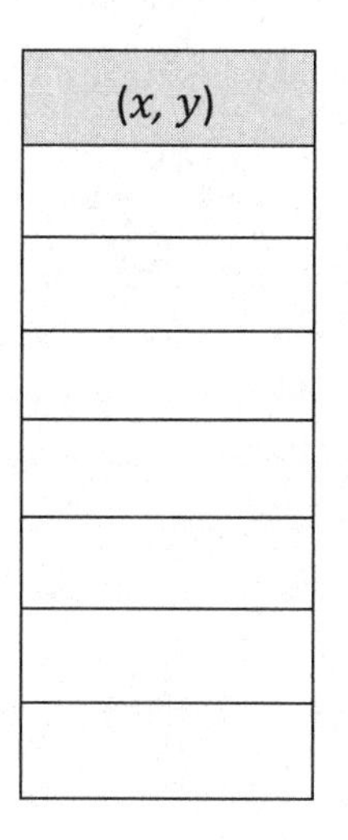

(x, y)

   c. Complete the drawing to create a figure that is symmetric about line s. For each point in Table A, record the symmetric point on the other side of s.

   d. Compare the y-coordinates in Table A with those in Table B. What do you notice?

   e. Compare the x-coordinates in Table A with those in Table B. What do you notice?

2. Use the plane to the right to complete the following tasks.

   a. Draw a line p whose rule is, *y is equal to x*.

   b. Plot the points from Table A on the grid in order. Then, draw line segments to connect the points.

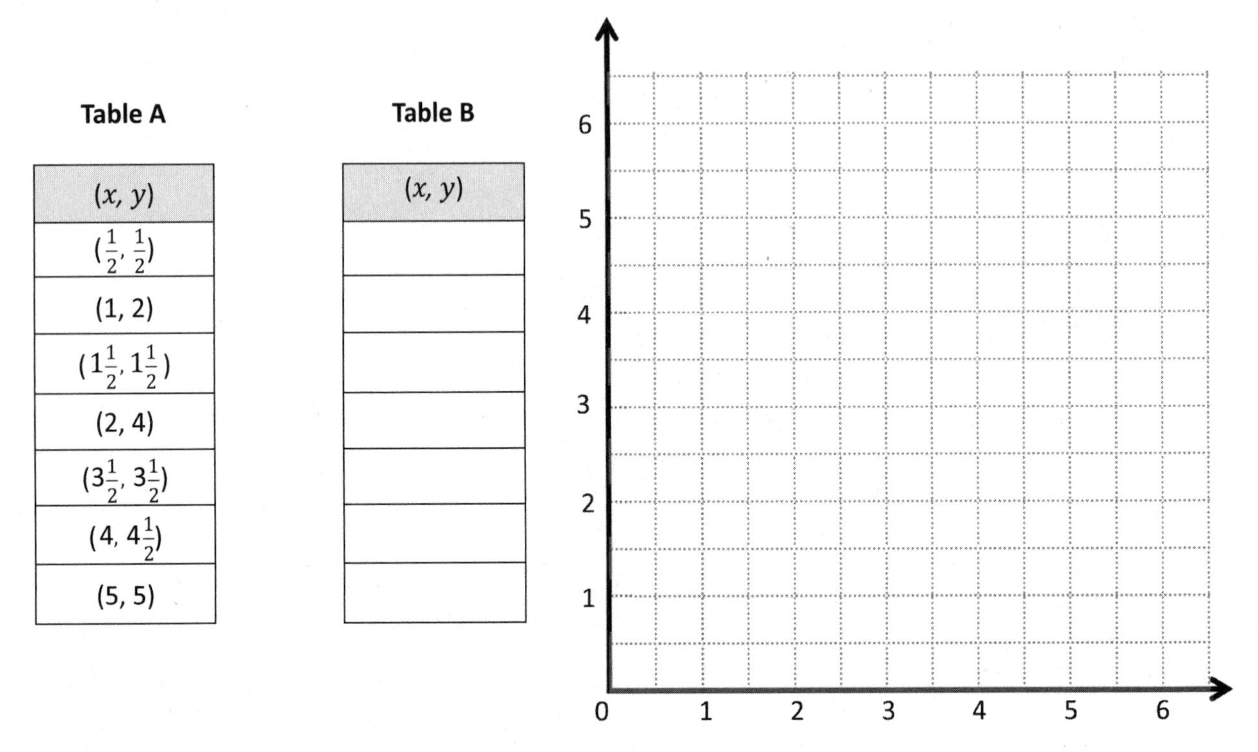

Table A

(x, y)
$(\frac{1}{2}, \frac{1}{2})$
$(1, 2)$
$(1\frac{1}{2}, 1\frac{1}{2})$
$(2, 4)$
$(3\frac{1}{2}, 3\frac{1}{2})$
$(4, 4\frac{1}{2})$
$(5, 5)$

Table B

(x, y)

   c. Complete the drawing to create a figure that is symmetric about line p. For each point in Table A, record the symmetric point on the other side of the line p in Table B.

   d. Compare the y-coordinates in Table A with those in Table B. What do you notice?

   e. Compare the x-coordinates in Table A with those in Table B. What do you notice?

The line graph below tracks the balance of Sheldon's checking account at the end of each day between June 10 and June 24. Use the information in the graph to answer the questions that follow.

I know that it is important to read the scale on the vertical axis so that I know what units the data is referring to. In this graph, the 1 means \$1,000, and the 2 means \$2,000. I can tell that each grid line skip-counts by \$250.

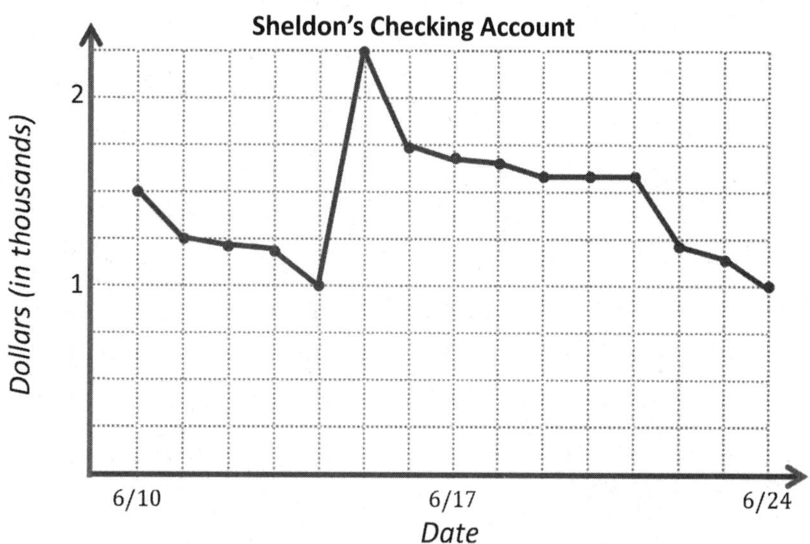

a. About how much money does Sheldon have in his checking account on June 10?

Sheldon has \$1,500 in his account on June 10. I can tell because the point is on the line exactly between \$1,000 and \$2,000.

b. If Sheldon spends \$250 from his checking account on June 24, about how much money will he have left in his account?

Sheldon will have \$750 left.

$\$1,000 - \$250 = \$750$

c. Sheldon received a payment from his job that went directly into his checking account. On which day did this most likely occur? Explain how you know.

The amount of money in his account increased by \$1,250 on June 15. This is most likely the day he was paid by his job.

d. Sheldon paid rent for his apartment during the time shown in the graph. On which day did this most likely occur? Explain how you know.

Sheldon might have paid his rent on either June 15 or June 21. These are the two days where Sheldon's account went down most quickly.

Name ______________________________ Date ______________

1. The line graph below tracks the balance of Howard's checking account, at the end of each day, between May 12 and May 26. Use the information in the graph to answer the questions that follow.

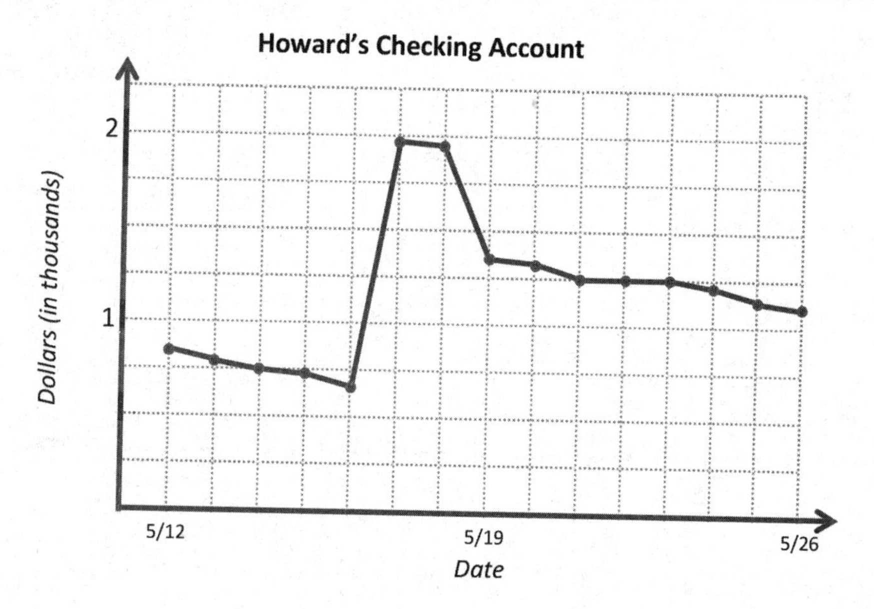

a. About how much money does Howard have in his checking account on May 21?

b. If Howard spends $250 from his checking account on May 26, about how much money will he have left in his account?

c. Explain what happened with Howard's money between May 21 and May 23.

d. Howard received a payment from his job that went directly into his checking account. On which day did this most likely occur? Explain how you know.

e. Howard bought a new television during the time shown in the graph. On which day did this most likely occur? Explain how you know.

2. The line graph below tracks Santino's time at the beginning and end of each part of a triathlon. Use the information in the graph to answer the questions that follow.

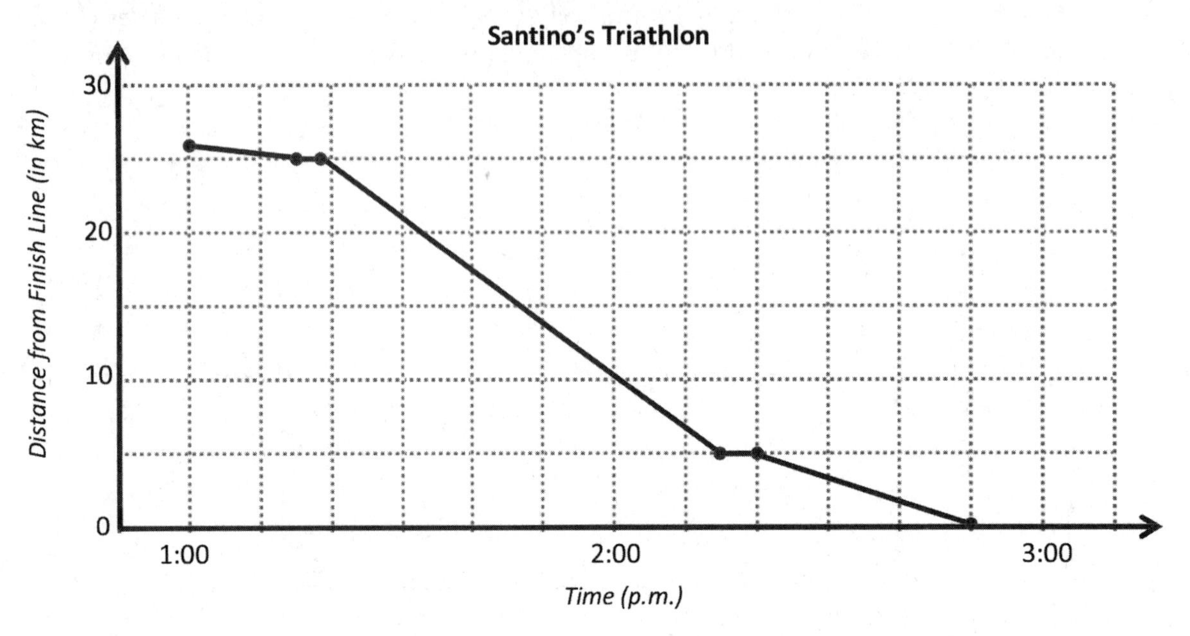

a. How long does it take Santino to finish the triathlon?

b. To complete the triathlon, Santino first swims across a lake, then bikes through the city, and finishes by running around the lake. According to the graph, what was the distance of the running portion of the race?

c. During the race, Santino pauses to put on his biking shoes and helmet and then later to change into his running shoes. At what times did this most likely occur? Explain how you know.

d. Which part of the race does Santino finish most quickly? How do you know?

e. During which part of the triathlon is Santino racing most quickly? Explain how you know.

Use the graph to answer the questions.

Hector left his home at 6:00 a.m. to train for a bicycle race. He used his GPS watch to keep track of the number of miles he traveled at the end of each hour of his trip. He uploaded the data to his computer, which gave him the line graph below:

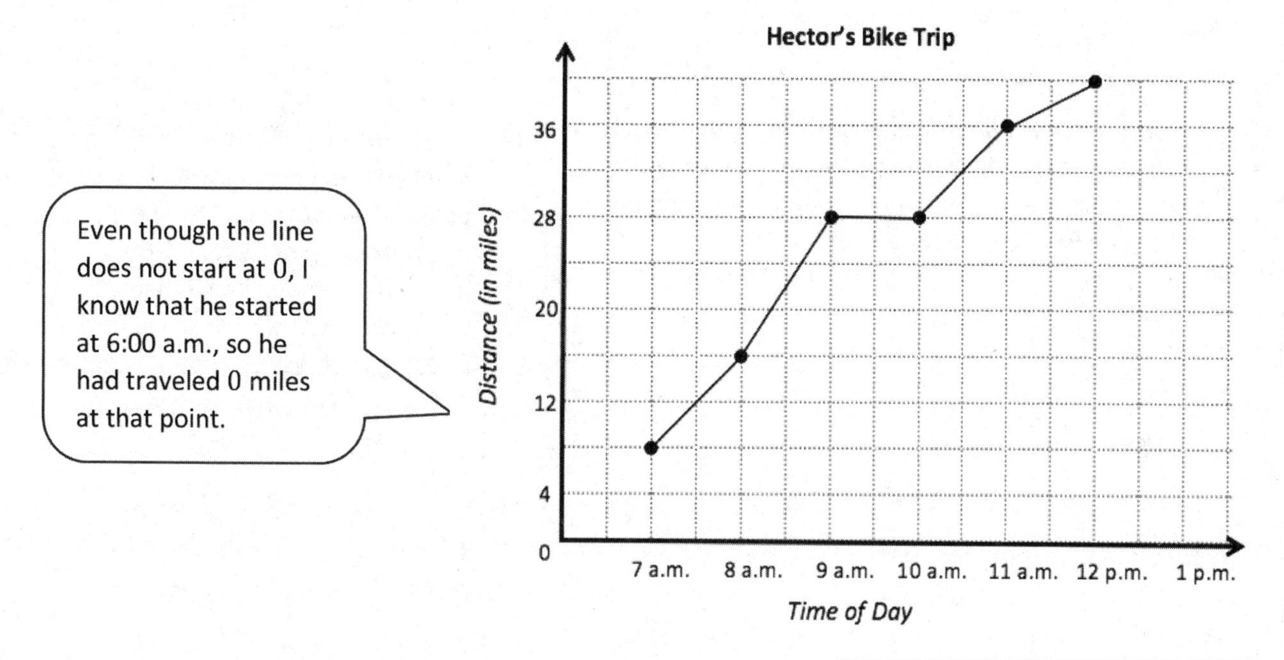

a. How far did Hector travel in all? How long did it take?

Hector traveled 40 miles in 6 hours.

Hector started at 6:00 a.m. and stopped at noon. That's 6 hours.

The last data point at 12:00 p.m. shows 40 miles.

b. Hector took a one-hour break to have a snack and take some pictures. What time did he stop? How do you know?

Hector took his break from 9 a.m. to 10 a.m. The horizontal line at this time tells me that Hector's distance did not change; therefore, he wasn't biking for that hour.

c. During which hour did Hector ride the slowest?

Hector's slowest hour was his last one between 11:00 a.m. and noon. He only rode 4 miles in that last hour whereas in the other hours he rode at least 8 miles (except when he took his break).

I also know I can look at how steep the line is between two points to help me know how fast or slow Hector rode. The line is not very steep between 11:00 a.m and noon, so I know that was his slowest hour.

Name ______________________________ Date ______________

Use the graph to answer the questions.

Johnny left his home at 6 a.m. and kept track of the number of kilometers he traveled at the end of each hour of his trip. He recorded the data in a line graph.

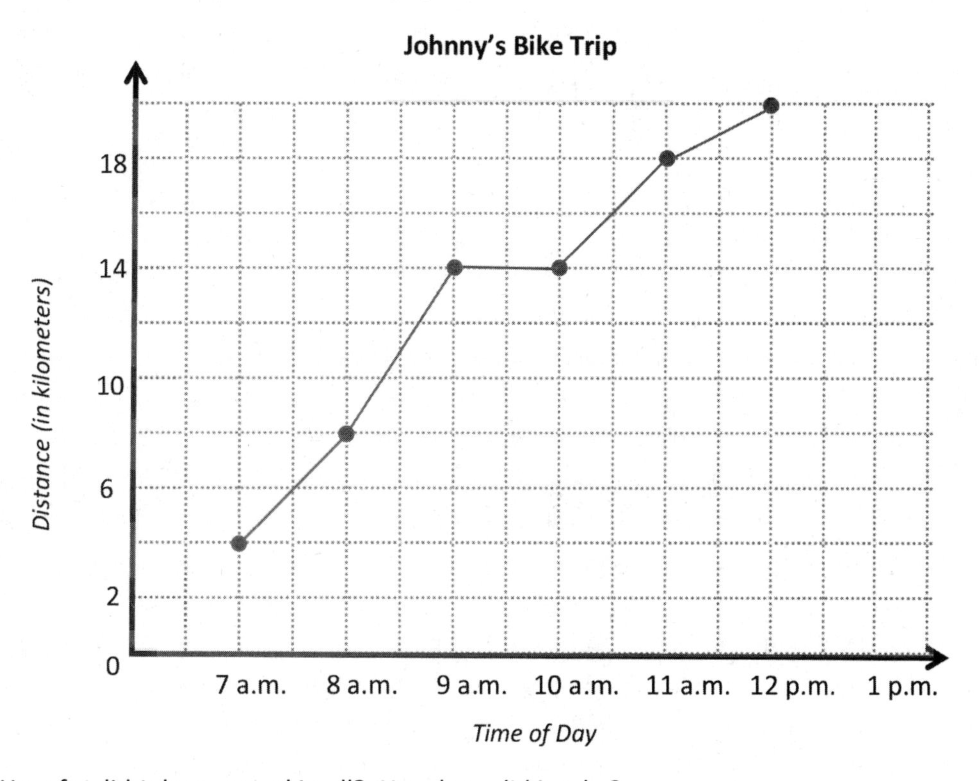

a. How far did Johnny travel in all? How long did it take?

b. Johnny took a one-hour break to have a snack and take some pictures. What time did he stop? How do you know?

c. Did Johnny cover more distance before his break or after? Explain.

d. Between which two hours did Johnny ride 4 kilometers?

e. During which hour did Johnny ride the fastest? Explain how you know.

Meyer read four times as many books as Zenin. Lenox read as many as Meyer and Zenin combined. Parks read half as many books as Zenin. In total, all four read 147 books. How many books did each child read?

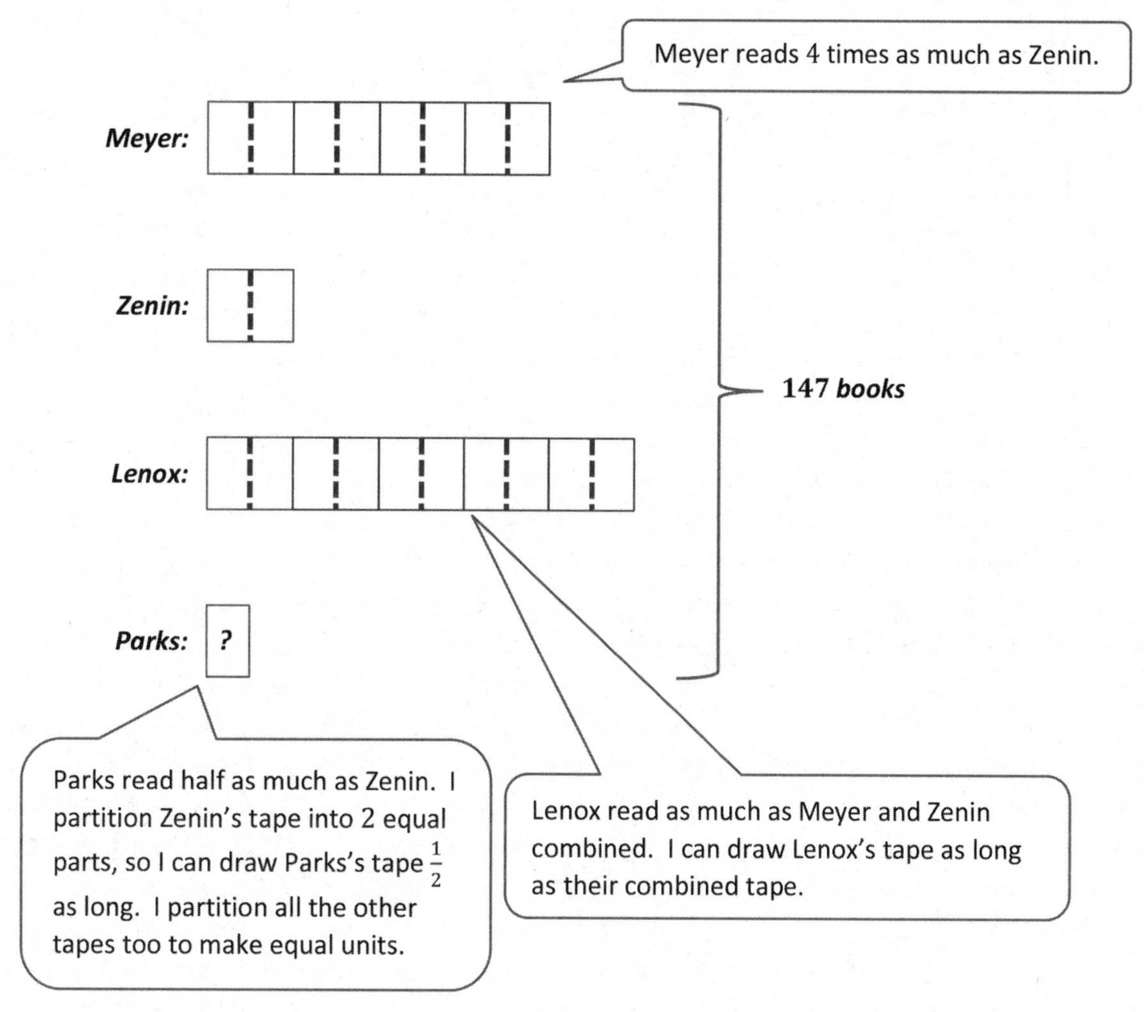

$21 \textit{ units} = 147 \textit{ books}$

$1 \textit{ unit} = 147 \textit{ books} \div 21 = 7 \textit{ books}$

	Parks read 7 books.
$7 \times 8 = 56$	*Meyer read 56 books.*
$7 \times 2 = 14$	*Zenin read 14 books.*
$56 + 14 = 70$	*Lenox read 70 books.*

Name ______________________________ Date ______________

1. Sara travels twice as far as Eli when going to camp. Ashley travels as far as Sara and Eli together. Hazel travels 3 times as far as Sara. In total, all four travel 888 miles to camp. How far does each of them travel?

The following problem is a brainteaser for your enjoyment. It is intended to encourage working together and family problem-solving fun. It is not a required element of this homework assignment.

2. A man wants to take a goat, a bag of cabbage, and a wolf over to an island. His boat will only hold him and one animal or item. If the goat is left with the cabbage, he'll eat it. If the wolf is left with the goat, he'll eat it. How can the man transport all three to the island without anything being eaten?

Solve using any method. Show all your thinking.

I know that squares have all 4 sides of equal length.

Study this diagram showing all the squares. Fill in the table.

Figure	Area in Square Centimeters
1	9 cm^2
2	$\mathbf{81\ cm^2}$
3	$\mathbf{36\ cm^2}$
5	$\mathbf{9\ cm^2}$
6	$\mathbf{9\ cm^2}$

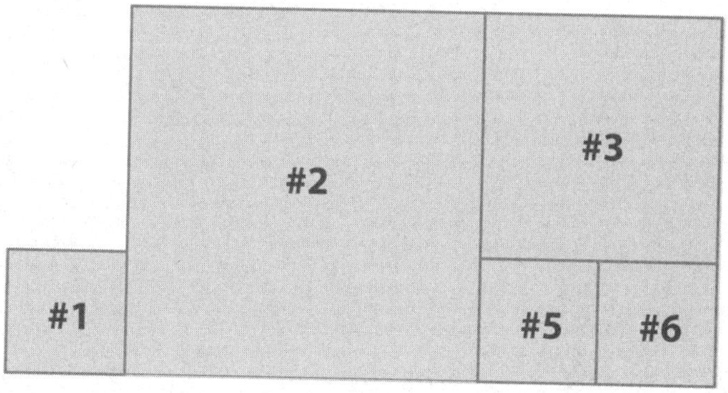

#2
#3
#1
#5
#6

The table says the area of Figure 1 is 9 cm^2.
$3 \text{ cm} \times 3 \text{ cm} = 9 \text{ cm}^2$
I know that each side of Figure 1 is 3 cm long.

Figures 5 and 6 are the same size as Figure 1. They also have an area of 9 cm^2.

Figure 3:

$\mathbf{3\ cm + 3\ cm = 6\ cm}$

$\mathbf{6\ cm \times 6\ cm = 36\ cm^2}$

Figure 3 shares a side with Figures 5 and 6. Since the side lengths of Figures 5 and 6 are 3 cm each, the side length of Figure 3 must be 6 cm.

Figure 2:

$\mathbf{6\ cm + 3\ cm = 9\ cm}$

$\mathbf{9\ cm \times 9\ cm = 81\ cm^2}$

Figure 2 shares a side with Figures 3 and 5. Since the side lengths of Figures 3 and 5 are 6 cm and 3 cm, respectively, the side length of Figure 2 must be 9 cm.

Name ______________________________ Date ______________

Solve using any method. Show all your thinking.

1. Study this diagram showing all the squares. Fill in the table.

Figure	Area in Square Feet
1	1 ft^2
2	
3	
4	9 ft^2
5	
6	1 ft^2
7	
8	
9	
10	
11	

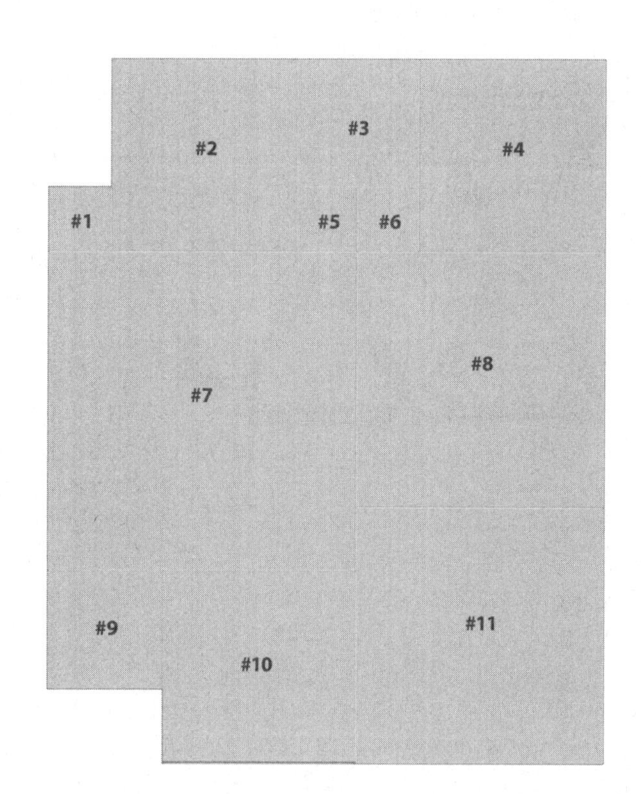

The following problem is a brainteaser for your enjoyment. It is intended to encourage working together and family problem-solving fun. It is not a required element of this homework assignment.

2. Remove 3 matches to leave 3 triangles.

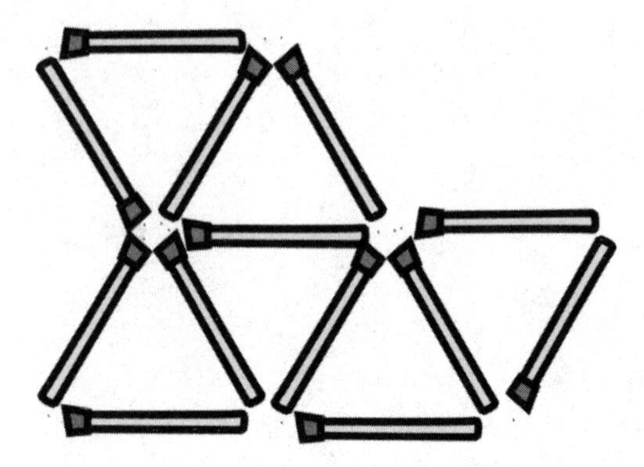

In the diagram, the length of Figure B is $\frac{4}{7}$ the length of Figure A. Figure A has an area of 182 in^2. Find the perimeter of the entire figure.

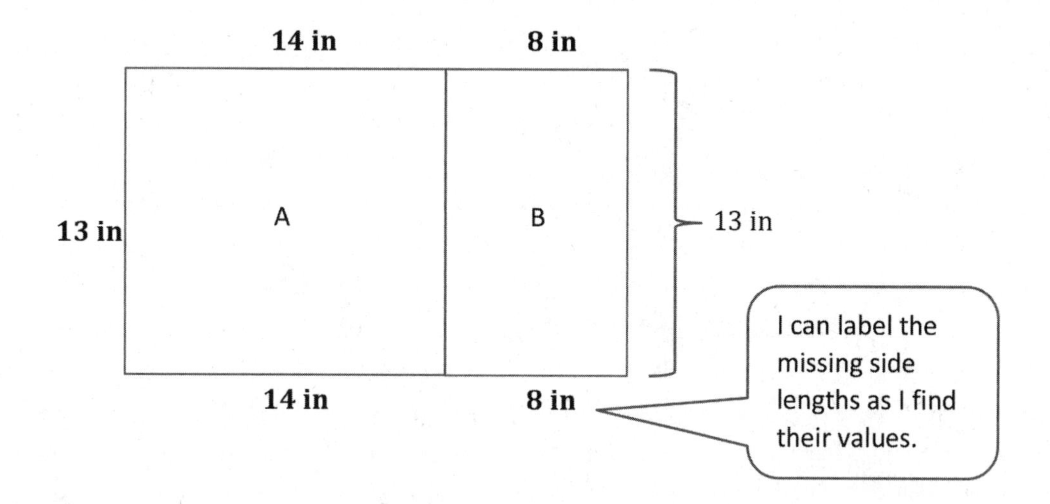

I can find the length of Figure A by dividing the area by the width.

Now that I know the length of Figure A, I can use it to find the length of Figure B.

I can find the perimeter of the entire figure by adding up all of the sides.

Figure A:

$$\text{Area} = \text{length} \times \text{width}$$
$$182 = \_\_\_ \times 13$$
$$182 \div 13 = 14$$

The length of Figure A is 14 inches.

Figure B:

$\frac{4}{7}$ *of 14 inches*

$$\frac{4}{7} \times 14$$
$$= \frac{4 \times 14}{7}$$
$$= \frac{56}{7}$$
$$= 8$$

The length of Figure B is 8 inches.

Entire Figure:

$$14 + 8 + 13 + 8 + 14 + 13 = 70$$

The perimeter of the entire figure is 70 inches.

Name ______________________________ Date ________________

1. In the diagram, the length of Figure S is $\frac{2}{3}$ the length of Figure T. If S has an area of 368 cm^2, find the perimeter of the figure.

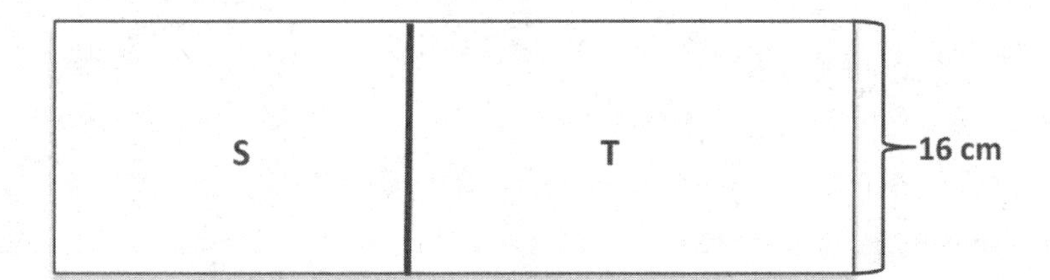

The following problems are puzzles for your enjoyment. They are intended to encourage working together and family problem-solving fun and are not a required element of this homework assignment.

2. Take 12 matchsticks arranged in a grid as shown below, and remove 2 matchsticks so 2 squares remain. How can you do this? Draw the new arrangement.

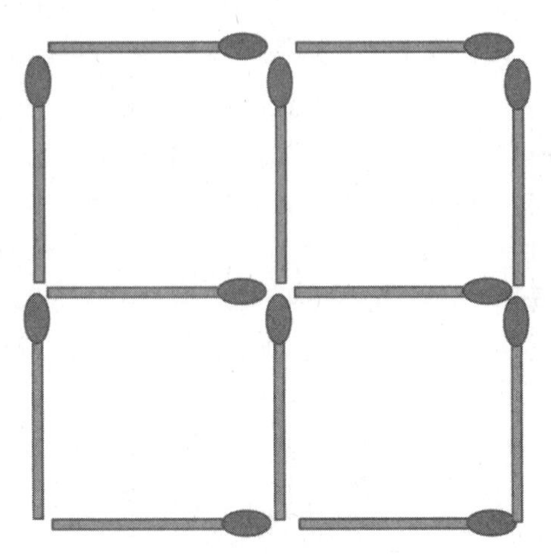

3. Moving only 3 matchsticks makes the fish turn around and swim the opposite way. Which matchsticks did you move? Draw the new shape.

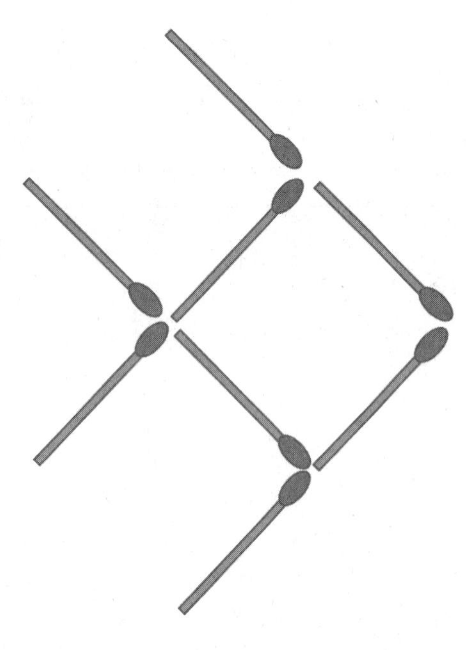

Howard's Baseball Camp welcomed 96 athletes on the first day of camp. Five-eighths of the athletes began practicing hitting. The hitting coach sent $\frac{2}{5}$ of the hitters to work on bunting. Half of the bunters were left-handed hitters. The left-handed bunters were put into teams of 2 to practice together. How many teams of 2 were practicing bunting?

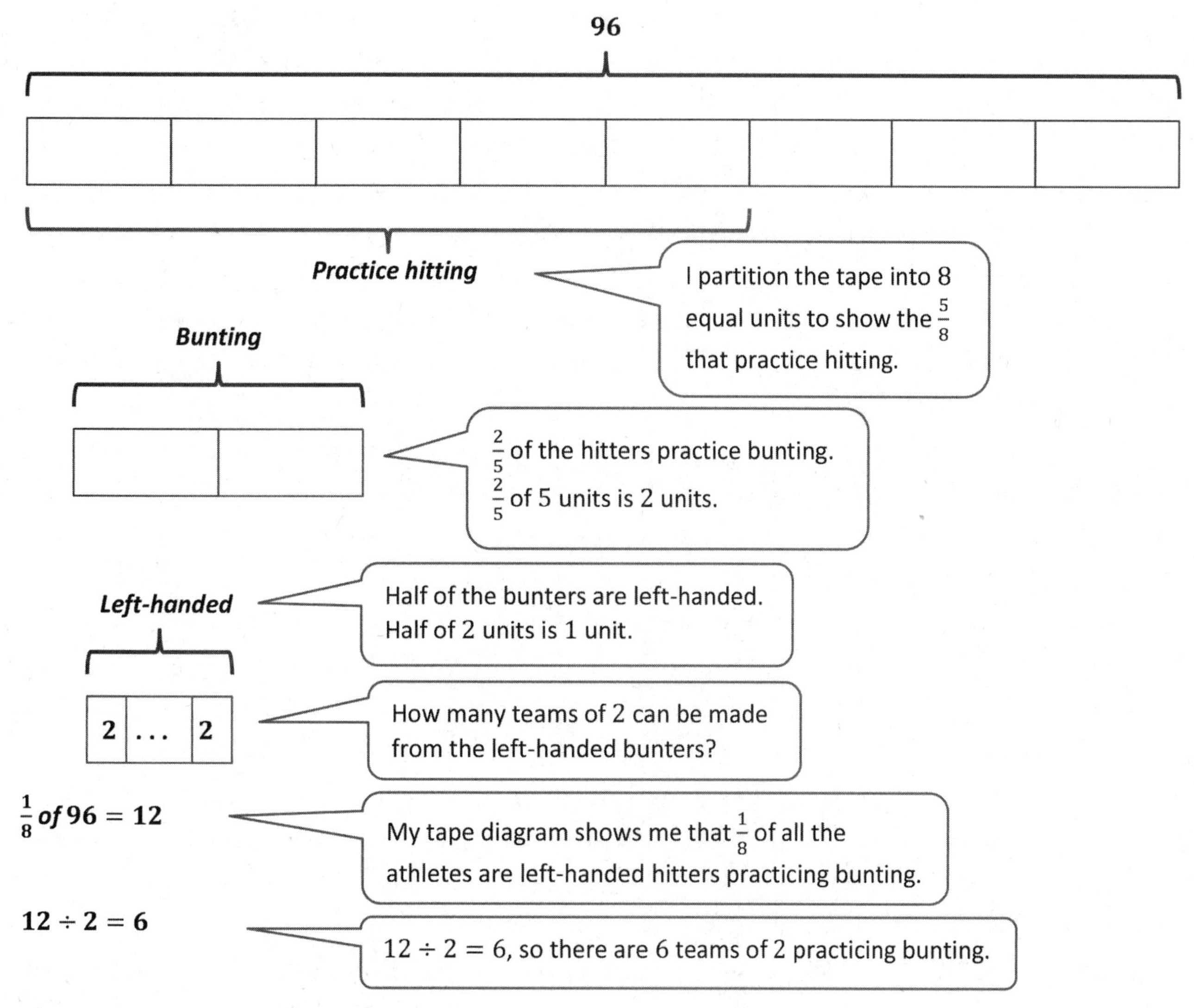

$\frac{1}{8}$ *of* $96 = 12$

My tape diagram shows me that $\frac{1}{8}$ of all the athletes are left-handed hitters practicing bunting.

$12 \div 2 = 6$

$12 \div 2 = 6$, so there are 6 teams of 2 practicing bunting.

There are 6 teams of 2 practicing bunting.

Name ______________________________ Date ________________

1. Pat's Potato Farm grew 490 pounds of potatoes. Pat delivered $\frac{3}{7}$ of the potatoes to a vegetable stand. The owner of the vegetable stand delivered $\frac{2}{3}$ of the potatoes he bought to a local grocery store, which packaged half of the potatoes that were delivered into 5-pound bags. How many 5-pound bags did the grocery store package?

The following problems are for your enjoyment. They are intended to encourage working together and family problem-solving fun. They are not a required element of this homework assignment.

2. Six matchsticks are arranged into an equilateral triangle. How can you arrange them into 4 equilateral triangles without breaking or overlapping any of them? Draw the new shape.

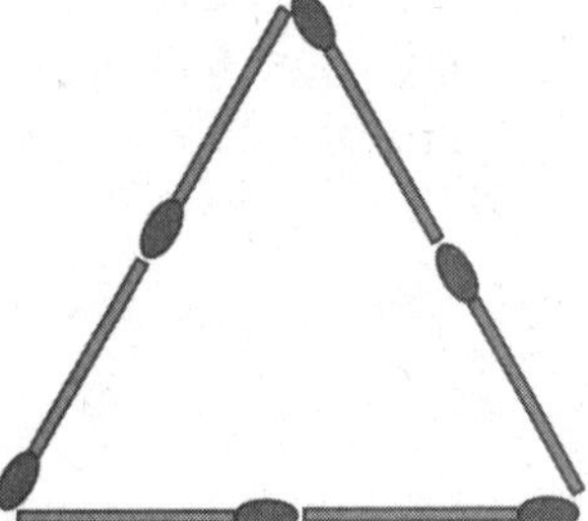

3. Kenny's dog, Charlie, is really smart! Last week, Charlie buried 7 bones in all. He buried them in 5 straight lines and put 3 bones in each line. How is this possible? Sketch how Charlie buried the bones.

Jason and Selena had \$96 altogether at first. After Jason spent $\frac{1}{5}$ of his money and Selena lent \$15 of her money, they had the same amount of money left. How much money did each of them have at first?

This is important. *After* Jason spends and Selena lends, *then* they have the same amount left. I need to make sure that my model shows this.

I partition the tape representing Jason's money into 5 equal parts to show the $\frac{1}{5}$ that he spent.

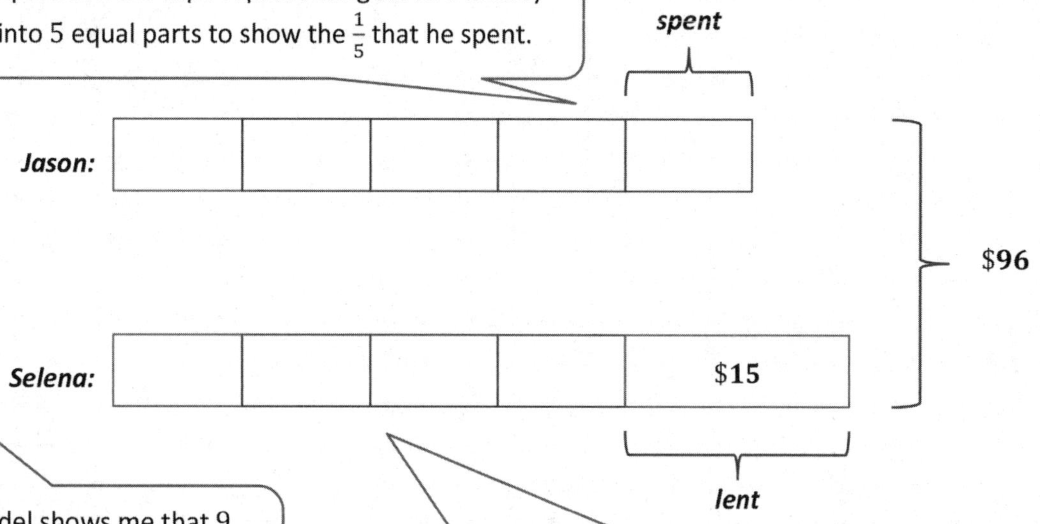

spent

Jason:

\$96

Selena: \$15

lent

My model shows me that 9 units, plus the \$15 that Selena lent, is equal to \$96.

To show that Selena and Jason have the same amount of money left, I partition the tape representing Selena's money the same way that I did Jason's.

$9 \textit{ units} + \$15 = \$96$

$9 \textit{ units} = \$81$

$1 \textit{ unit} = \$81 \div 9 = \$9$

Now that I know the value of 1 unit, I can find out how much money they each had at first.

Jason:

$1 \textit{ unit} = \$9$

$5 \textit{ units} = 5 \times \$9 = \$45$

Jason had \$45 at first.

Selena:

$1 \textit{ unit} = \$9$

$4 \textit{ units} = 4 \times \$9 = \$36$

$\$36 + \$15 = \$51$

Selena had \$51 at first.

Name ______________________________ Date ______________

1. Fred and Ethyl had 132 flowers altogether at first. After Fred sold $\frac{1}{4}$ of his flowers and Ethyl sold 48 of her flowers, they had the same number of flowers left. How many flowers did each of them have at first?

The following problems are puzzles for your enjoyment. They are intended to encourage working together and family problem-solving fun. They are not a required element of this homework assignment.

2. Without removing any, move 2 matchsticks to make 4 identical squares. Which matchsticks did you move? Draw the new shape.

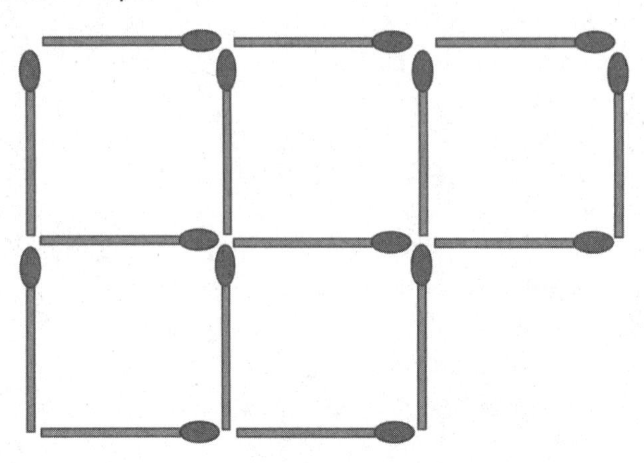

3. Move 3 matchsticks to form exactly (and only) 3 identical squares. Which matchsticks did you move? Draw the new shape.

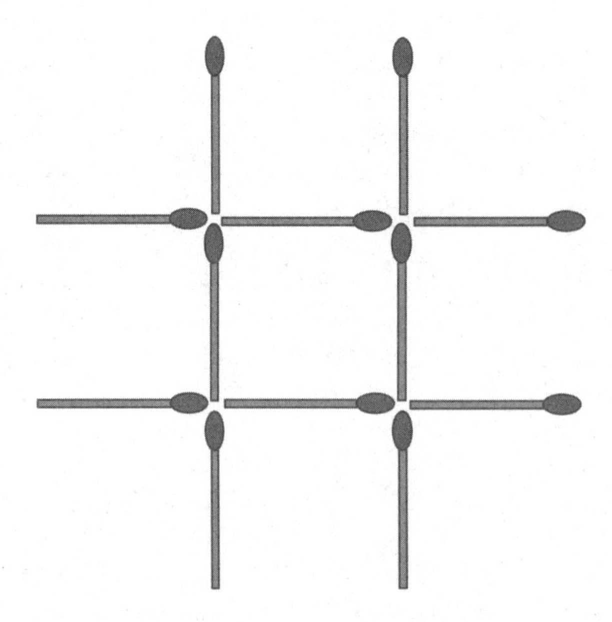

1. For the phrase below, write a numerical expression, and then evaluate your expression.

   Subtract three halves from one sixth of forty-two.

$\frac{1}{6} \times 42 - \frac{3}{2}$

$= \frac{42}{6} - \frac{3}{2}$

$= 7 - \frac{3}{2}$

$= 7 - 1\frac{1}{2}$

$= 5\frac{1}{2}$

Even though it says the word "*subtract*" first, I need to have something to subtract from. So I won't subtract until I find the value of "*one sixth of forty-two*."

2. Write at least 2 numerical expressions for the phrase below. Then, solve.

   Two fifths of nine

$\frac{2}{5} \times 9$ $\quad\quad$ $\left(\frac{1}{5} \times 9\right) \times 2$

This is "*one fifth of nine, doubled,*" which is equal to "*two fifths of nine.*"

$\frac{2}{5} \times 9$

$= \frac{2 \times 9}{5}$

$= \frac{18}{5}$

$= 3\frac{3}{5}$

"*Two fifths of nine*" is equal to $3\frac{3}{5}$.

3. Use $<$, $>$, or $=$ to make true number sentences without calculating. Explain your thinking.

a. $\left(481 \times \frac{9}{16}\right) \times \frac{2}{10}$ $<$ $\left(481 \times \frac{9}{16}\right) \times \frac{7}{10}$

Both expressions have the same first factor, $\left(481 \times \frac{9}{16}\right)$.

Since the second factor, $\frac{7}{10}$, *is greater than* $\frac{2}{10}$, *the expression on the right is greater.*

b. $\left(4 \times \frac{1}{10}\right) + \left(9 \times \frac{1}{100}\right)$ 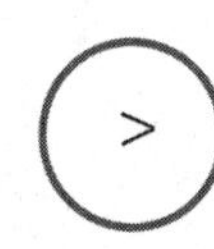 0.409

The expression on the left is equal to 0.49.

The expression on the right also has 0 *ones and* 4 *tenths, but there are* 0 *hundredths in* 0.409.

Name ______________________________ Date ______________

1. For each written phrase, write a numerical expression, and then evaluate your expression.

a. Forty times the sum of forty-three and fifty-seven

Numerical expression:

Solution:

b. Divide the difference between one thousand three hundred and nine hundred fifty by four.

Numerical expression:

Solution:

c. Seven times the quotient of five and seven

Numerical expression:

Solution:

d. One fourth the difference of four sixths and three twelfths

Numerical expression:

Solution:

2. Write at least 2 numerical expressions for each written phrase below. Then, solve.

   a. Three fifths of seven

   b. One sixth the product of four and eight

3. Use <, >, or = to make true number sentences without calculating. Explain your thinking.

   a. 4 tenths + 3 tens + 1 thousandth ◯ 30.41

   b. $\left(5 \times \frac{1}{10}\right) + \left(7 \times \frac{1}{1000}\right)$ ◯ 0.507

   c. 8 × 7.20 ◯ 8 × 4.36 + 8 × 3.59

1. Use the RDW process to solve the word problem below.

   Daquan brought 32 cupcakes to school. Of those cupcakes, $\frac{3}{4}$ were chocolate, and the rest were vanilla. Daquan's classmates ate $\frac{5}{8}$ of the chocolate cupcakes and $\frac{3}{4}$ of the vanilla. How many cupcakes are left?

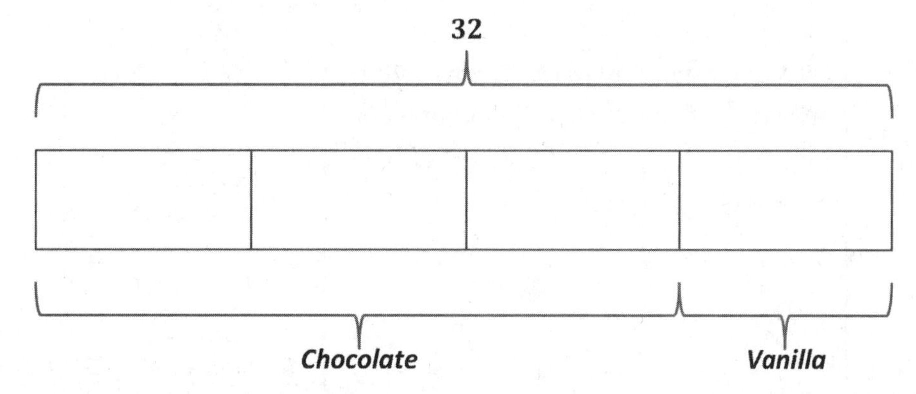

(of which $\frac{5}{8}$ are eaten)

(of which $\frac{3}{4}$ are eaten)

Of all the cupcakes, 24 are chocolate.

Chocolate eaten:

$\frac{3}{4}$ *of* $32 = \frac{3 \times 32}{4} = \frac{96}{4} = 24$

$\frac{5}{8}$ *of* $24 = \frac{5 \times 24}{8} = \frac{120}{8} = 15$

Of the 24 chocolate cupcakes, 15 were eaten.

15 *chocolate cupcakes were eaten.*

Of all the cupcakes, 8 are vanilla.

Vanilla eaten:

$\frac{1}{4}$ *of* $32 = \frac{1 \times 32}{4} = \frac{32}{4} = 8$

$\frac{3}{4}$ *of* $8 = \frac{3 \times 8}{4} = \frac{24}{4} = 6$

Of the 8 vanilla cupcakes, 6 were eaten.

6 *vanilla cupcakes were eaten.*

Cupcakes left:

$32 - (15 + 6) = 32 - 21 = 11$

11 *cupcakes are left.*

I find the number of leftover cupcakes by subtracting those that were eaten from the 32 original cupcakes.

2. Write and solve a word problem for the expression in the chart below.

Expression	Word Problem	Solution
$5 - \left(\frac{5}{12} + \frac{1}{3}\right)$	***During her 5-day work week, Mrs. Gomez spends $\frac{5}{12}$ of one day and $\frac{1}{3}$ of another in meetings. How much of her work week is <u>not</u> spent in meetings?***	$5 - \left(\frac{5}{12} + \frac{1}{3}\right)$ $= 5 - \left(\frac{5}{12} + \frac{4}{12}\right)$ $= 5 - \frac{9}{12}$ $= 4\frac{3}{12}$ $= 4\frac{1}{4}$ ***$4\frac{1}{4}$ days of Mrs. Gomez' work week was not spent in meetings.***

Name ______________________________ Date ____________

1. Use the RDW process to solve the word problems below.

   a. There are 36 students in Mr. Meyer's class. Of those students, $\frac{5}{12}$ played tag at recess, $\frac{1}{3}$ played kickball, and the rest played basketball. How many students in Mr. Meyer's class played basketball?

   b. Julie brought 24 apples to school to share with her classmates. Of those apples, $\frac{2}{3}$ are red, and the rest are green. Julie's classmates ate $\frac{3}{4}$ of the red apples and $\frac{1}{2}$ of the green apples. How many apples are left?

2. Write and solve a word problem for each expression in the chart below.

Expression	Word Problem	Solution
$144 \times \frac{7}{12}$		
$9 - \left(\frac{4}{9} + \frac{1}{3}\right)$		
$\frac{3}{4} \times (36 + 12)$		

Name ______________________________ Date ______________

1. Use what you learned about your fluency skills today to answer the questions below.

   a. Which skills should you practice this summer to maintain and build your fluency? Why?

   b. Write a goal for yourself about a skill that you want to work on this summer.

   c. Explain the steps you can take to reach your goal.

   d. How will reaching this goal help you as a math student?

2. In the chart below, plan a new fluency activity that you can play at home this summer to help you build or maintain a skill that you listed in Problem 1(a). When planning your activity, be sure to think about the factors listed below:

- The materials that you'll need.
- Who can play with you (if more than 1 player is needed).
- The usefulness of the activity for building your skills.

Skill:
Name of Activity:
Materials Needed:
Description:

Use your ruler, protractor, and set square to help you give as many names as possible for each figure below. Then, explain your reasoning for how you named each figure.

Figure	Names	Reasoning for Names
a.	*quadrilateral* *trapezoid*	***This figure is a <u>quadrilateral</u> because it is a closed figure with 4 sides.*** ***It's also a <u>trapezoid</u> because it has at least one pair of parallel sides. The top and bottom sides are parallel.***
b. I use my protractor and ruler to measure the angles and the side lengths. This shape has four 90° angles and four equal sides. That means it's a square, but it has other names, too.	*quadrilateral* *trapezoid* *parallelogram* *rectangle* *rhombus* *kite* *square*	***This figure is a <u>quadrilateral</u> because it is a closed figure with 4 sides.*** ***It's also a <u>trapezoid</u> because it has at least one pair of parallel sides. This shape actually has 2 pairs.*** ***This shape is also a <u>parallelogram</u> because opposite sides are both parallel and equal in length.*** ***It's also a <u>rectangle</u> because it has 4 right angles.*** ***It's a <u>rhombus</u> because all 4 sides are equal in length.*** ***It's also a <u>kite</u> because it has 2 pairs of adjacent sides that are equal in length.*** ***But most specifically, it's a <u>square</u> because it has 4 right angles and 4 sides of equal length.***

Name ______________________________ Date ______________

1. Use your ruler, protractor, and set square to help you give as many names as possible for each figure below. Then, explain your reasoning for how you named each figure.

Figure	Names	Reasoning for Names
a.		
b.		
c.		
d.		

2. Mark draws a figure that has the following characteristics:

   - Exactly 4 sides that are each 7 centimeters long.
   - Two sets of parallel lines.
   - Exactly 4 angles that measure 35 degrees, 145 degrees, 35 degrees, and 145 degrees.

   a. Draw and label Mark's figure below.

   b. Give as many names of quadrilaterals as possible for Mark's figure. Explain your reasoning for the names of Mark's figure.

   c. List the names of Mark's figure in Problem 2(b) in order from least specific to most specific. Explain your thinking.

Name ______________________________ Date ________________

Teach someone at home how to play one of the games you played today with your pictorial vocabulary cards. Then, answer the questions below.

1. What games did you play?

2. Who played the games with you?

3. What was it like to teach someone at home how to play?

4. Did you have to teach the person who played with you any of the math concepts before you could play? Which ones? What was that like?

5. When you play these games at home again, what changes will you make? Why?

Lesson Notes

To get a better understanding of the Fibonacci numbers, watch the short video, "Doodling in Math: Spirals, Fibonacci, and Being a Plant" by Vi Hart (http://youtu.be/ahXIMUkSXXO).

1. In your own words, describe what you know about the Fibonacci numbers.

   The Fibonacci numbers are really interesting. They're a list of numbers. You can always find the next number in the series by adding together the 2 numbers that come before it.

   For example, if part of the series is 13 and then 21, then the next number in the list will be 34 because $13 + 21 = 34$.

   I can remember the first few Fibonacci numbers:

   $$1, 1, 2, 3, 5, 8, 13, 21, 34.$$

2. Describe what the drawing you did in class today looked like.

   I can visualize what we drew in class. It looked like this:

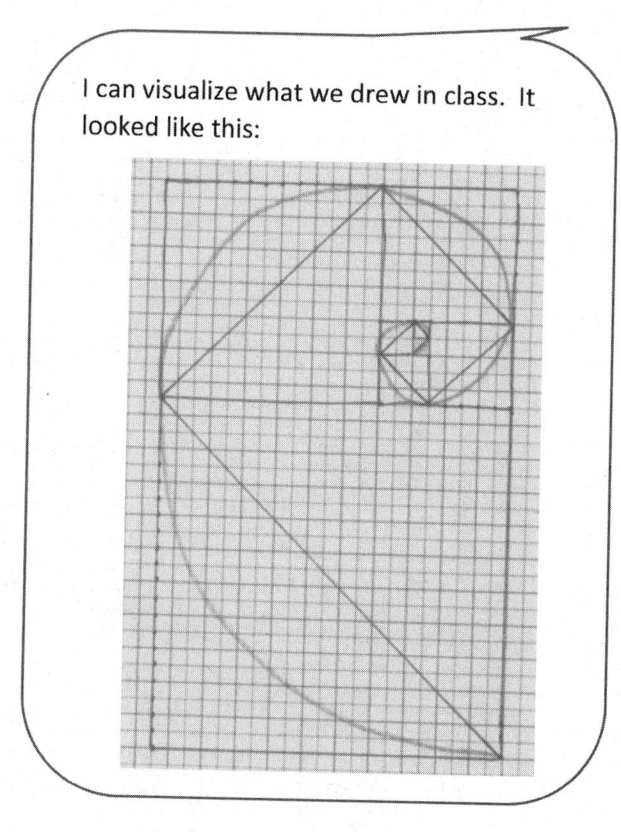

   At first, the drawing just looked like a bunch of square boxes drawn near one another that had a side in common. But then we drew a diagonal line across each square. Then we drew a more curved line inside each square, and it created this really neat spiral pattern, kind of like a seashell.

   After we drew it, we wrote down the side length of each square we drew and realized that they were the Fibonacci numbers. In other words, the first 2 squares we drew had a side length of 1, then the next square had a side length of 2, then 3, then 5, and so on.

Name ______________________________ Date ________________

1. List the Fibonacci numbers up to 21, and create, on the graph below, a spiral of squares corresponding to each of the numbers you write.

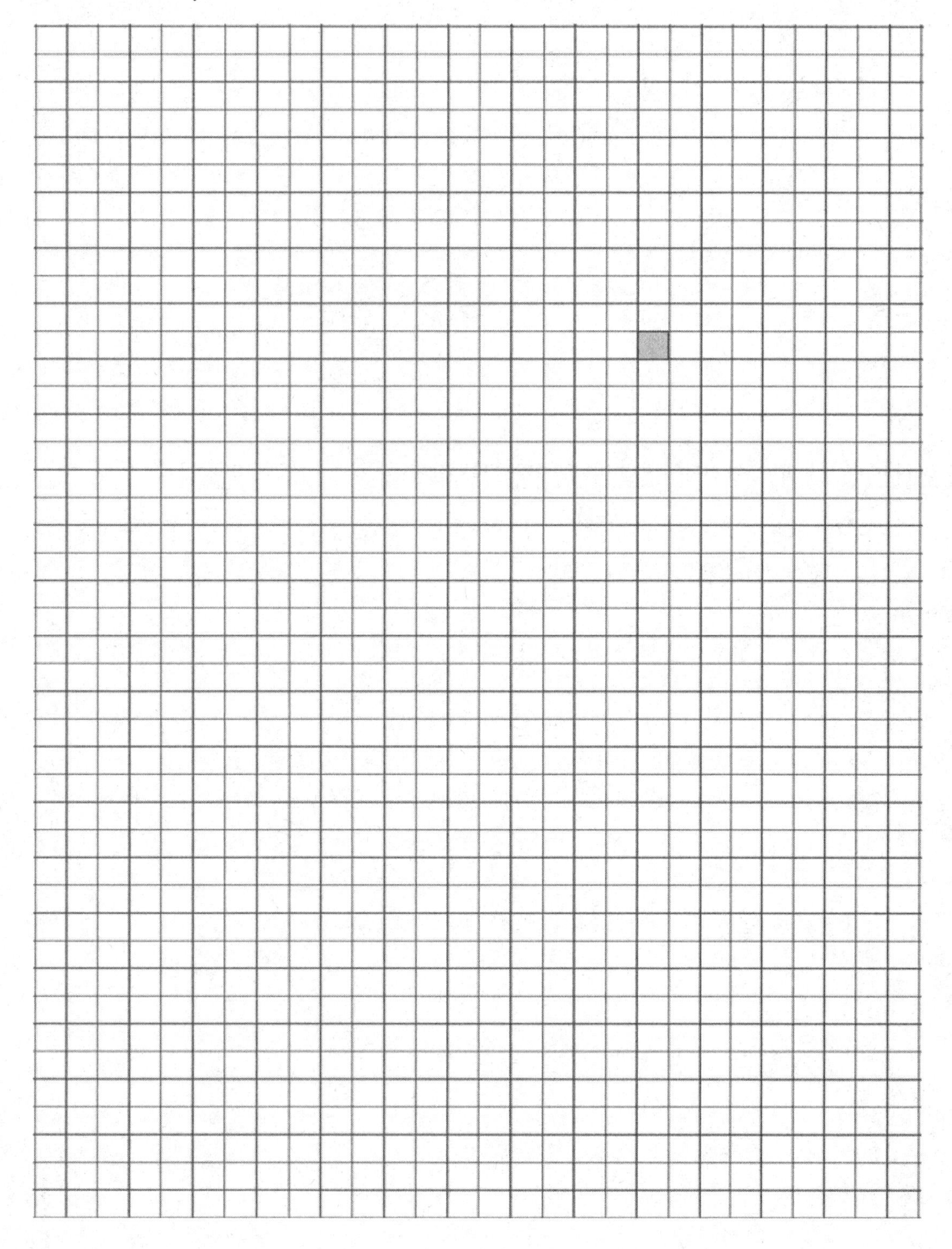

2. In the space below, write a rule that generates the Fibonacci sequence.

3. Write at least the first 15 numbers of the Fibonacci sequence.

Lesson Notes

To get a better understanding of the Fibonacci numbers, watch the short video, "Doodling in Math: Spirals, Fibonacci, and Being a Plant" by Vi Hart (http://youtu.be/ahXIMUkSXXO).

1. Complete the Fibonacci sequence in the table below.

The values in the top row tell the order of the numbers in the sequence. For example, this is the 6th number in the sequence.

1	2	3	4	5	6	7	8	9
1	1	2	3	**5**	**8**	**13**	**21**	**34**

I can find the value of the next number in the sequence by adding together the two previous numbers. $5 + 8 = 13$; therefore, the 7th number in the sequence is 13.

2. If the 12th and 13th numbers in the sequence are 144 and 233, respectively, what is the 11th number in the series?

$\_\_\_ + 144 = 233$

$233 - 144 = 89$

What number plus 144 is equal to 233? I can use subtraction to solve.

The 11th number in the series is 89.

Name ____________________ Date ____________

1. Jonas played with the Fibonacci sequence he learned in class. Complete the table he started.

1	2	3	4	5	6	7	8	9	10
1	1	2	3	5	8				

11	12	13	14	15	16	17	18	19	20

2. As he looked at the numbers, Jonas realized he could play with them. He took two consecutive numbers in the pattern and multiplied them by themselves and then added them together. He found they made another number in the pattern. For example, $(3 \times 3) + (2 \times 2) = 13$, another number in the pattern. Jonas said this was true for any two consecutive Fibonacci numbers. Was Jonas correct? Show your reasoning by giving at least two examples of why he was or was not correct.

3. Fibonacci numbers can be found in many places in nature, for example, the number of petals in a daisy, the number of spirals in a pine cone or a pineapple, and even the way branches grow on a tree. Find an example of something natural where you can see a Fibonacci number in action, and sketch it here.

Find a rectangular box at your home. Use a ruler to measure the dimensions of the box to the nearest centimeter. Then, calculate the volume of the box.

I find the volume of rectangular prisms, or boxes, by multiplying the 3 dimensions together.
$\text{Volume} = \text{length} \times \text{width} \times \text{height}$

Item	Length	Width	Height	Volume
Toy Shoe Box	**8 cm**	**3 cm**	**6 cm**	**144 cm^3**

The length of the shoe box was exactly 7.5 cm, but the directions said to measure to the nearest centimeter. I round 7.5 up to 8.

$8 \times 3 \times 6 = 24 \times 6 = 144$
The volume of the shoe box is 144 cubic centimeters.

Name____ Date____________________

1. Find various rectangular boxes at your home. Use a ruler to measure the dimensions of each box to the nearest centimeter. Then, calculate the volume of each box. The first one is partially done for you.

Item	Length	Width	Height	Volume
Juice Box	11 cm	2 cm	5 cm	

2. The dimensions of a small juice box are 11 cm by 4 cm by 7 cm. The super-size juice box has the same height of 11 cm but double the volume. Give two sets of the possible dimensions of the super-size juice box and the volume.

Credits

Great Minds® has made every effort to obtain permission for the reprinting of all copyrighted material. If any owner of copyrighted material is not acknowledged herein, please contact Great Minds for proper acknowledgment in all future editions and reprints of this module.